Competition Policy in Europe and
North America:
Economic Issues and Institutions

# FUNDAMENTALS OF PURE AND APPLIED ECONOMICS

## EDITORS IN CHIEF

J. LESOURNE, Conservatoire National des Arts et Métiers, Paris, France

H. SONNENSCHEIN, University of Pennsylvania, Philadelphia, PA, USA

## ADVISORY BOARD

K. ARROW, Stanford, CA, USA
W. BAUMOL, Princeton, NJ, USA
W. A. LEWIS, Princeton, NJ, USA
S. TSURU, Tokyo, Japan

*Fundamentals of Pure and Applied Economics* is an international series of titles divided by discipline into sections. A list of sections and their editors and of published titles may be found at the back of this volume.

# Competition Policy in Europe and North America: Economic Issues and Institutions

**W. S. Comanor**
*University of California, Santa Barbara, USA*

**K. George**
*University College of Swansea, UK*

**A. Jacquemin**
*Université Catholique de Louvain, Belgium*

**F. Jenny**
*E.S.S.E.C., Paris, France*

**E. Kantzenbach**
*Institut für wirtschaftsforschung, Hamburg, Federal Republic of Germany*

**J. A. Ordover**
*New York University, USA*

**L. Waverman**
*University of Toronto, Canada*

A volume in the Theory of the Firm and Industrial Organization section
edited by
**Alexis Jacquemin**
*Université Catholique de Louvain, Belgium*

harwood academic publishers
chur · london · paris · new york · melbourne

© 1990 by Harwood Academic Publishers GmbH
Poststrasse 22, 7000 Chur, Switzerland
All rights reserved

Harwood Academic Publishers

Post Office Box 197
London WC2E 9PX
United Kingdom

58, rue Lhomond
75005 Paris
France

Post Office Box 786
Cooper Station
New York, New York 10276
United States of America

Private Bag 8
Camberwell, Victoria 3124
Australia

**Library of Congress Cataloging-in-Publication Data**

Competition policy in Europe and North America : economic issues and institutions / W.S. Comanor . . . [et al.].
    p.    cm. — (Fundamentals of pure and applied economics, ISSN 0191-1708; v. 43. Theory of the firm and industrial organization section)
  Includes bibliographical references and index.
  ISBN 3-7186-5059-2
  1. Competition—Government policy—European Economic Community countries. 2. Competition—Government policy—United States. 3. Competition—Government policy—Canada. I. Series.
  HF1532.92.C66   1990
  338.6'048'094—dc20                                           90-44952
                                                                                           CIP

No part of this book may be reproduced or utilized in any form or by any means, electronic or mechanical, including photocopying and recording, or by any information storage or retrieval system, without permission in writing from the publisher. Printed in the United Kingdom.

# Contents

*Introduction to the Series* vii

INTRODUCTION: COMPETITION AND COMPETITION POLICY IN MARKET ECONOMIES
by Alexis Jacquemin 1

ECONOMIC FOUNDATIONS OF COMPETITION POLICY
by Janusz A. Ordover 7
1. Introduction 7
2. The Normative Economics of Competition and Monopoly 9
3. Indices of Market Power 25
4. Market Definition 35
5. Concluding Comments 38

UNITED STATES ANTITRUST POLICY: ISSUES AND INSTITUTIONS
by William S. Comanor 43
1. Introduction 43
2. A Brief Review 45
3. The Prohibition of Collusive Behavior 48
4. The Exercise and Abuse of Monopoly Power 53
5. Mergers Between Rivals 58
6. Mergers Between Buyers and Sellers 65
7. Vertical Contractual Relationships 67
8. Exclusive-Dealing Arrangements 71
9. Some Conclusions 72

CANADIAN COMPETITION LAW: 100 YEARS OF EXPERIMENTATION
by Leonard Waverman 73
1. Introduction and Summary 73
2. History of the Law 78
3. Collusive Activities 81
4. Monopoly 88
5. Mergers 93
6. Vertical Restraints 100

## UK COMPETITION POLICY: ISSUES AND INSTITUTIONS
by Ken George — 104
1. Introduction — 104
2. Horizontal Collusive Behaviour — 109
3. Monopoly and the Abuse of Market Power — 118
4. Mergers — 126
5. Vertical Relationships — 136

## FRENCH COMPETITION POLICY IN PERSPECTIVE
by F. Jenny — 146
1. The Emergence of Competition Law in France — 146
2. The Enforcement of Competition Law in France — 162
3. Conclusion — 187

## COMPETITION POLICY IN WEST GERMANY: A COMPARISON WITH THE ANTITRUST POLICY OF THE UNITED STATES
by Erhard Kantzenbach — 189
1. Historical Experiences — 189
2. Basic Philosophies — 190
3. Prohibition of Cartel Agreements — 191
4. Control of Dominant Enterprises — 193
5. Merger Control — 196
6. Horizontal Mergers — 197
7. Non-Horizontal Mergers — 200
8. Main Critics of the German Policy — 202

## COMPETITION POLICY IN THE EUROPEAN COMMUNITY
by Ken George and Alexis Jacquemin — 206
1. Introduction — 206
2. Horizontal Collusive Behaviour — 209
3. Vertical Arrangements — 223
4. Monopoly and the Abuse of Market Power — 228
5. Merger Policy — 233

## CONCLUSIONS FOR COMPETITION POLICY
by William S. Comanor, with Ken George, F. Jenny and Leonard Waverman — 246
1. Rules or Discretionary Authority — 248
2. Rule-Making for Competition Policy — 250
3. Prospects for an Effective Competition Policy — 252

*Index* — 253

# Introduction to the Series

Drawing on a personal network, an economist can still relatively easily stay well informed in the narrow field in which he works, but to keep up with the development of economics as a whole is a much more formidable challenge. Economists are confronted with difficulties associated with the rapid development of their discipline. There is a risk of "balkanization" in economics, which may not be favorable to its development.

*Fundamentals of Pure and Applied Economics* has been created to meet this problem. The discipline of economics has been subdivided into sections (listed at the back of this volume). These sections comprise short books, each surveying the state of the art in a given area.

Each book starts with the basic elements and goes as far as the most advanced results. Each should be useful to professors needing material for lectures, to graduate students looking for a global view of a particular subject, to professional economists wishing to keep up with the development of their science, and to researchers seeking convenient information on questions that incidentally appear in their work.

Each book is thus a presentation of the state of the art in a particular field rather than a step-by-step analysis of the development of the literature. Each is a high-level presentation but accessible to anyone with a solid background in economics, whether engaged in business, government, international organizations, teaching, or research in related fields.

Three aspects of *Fundamentals of Pure and Applied Economics* should be emphasized:

—First, the project covers the whole field of economics, not only theoretical or mathematical economics.
—Second, the project is open-ended and the number of books is not predetermined. If new and interesting areas appear, they will generate additional books.

—Last, all the books making up each section will later be grouped to constitute one or several volumes of an Encyclopedia of Economics.

The editors of the sections are outstanding economists who have selected as authors for the series some of the finest specialists in the world.

*J. Lesourne*                                              *H. Sonnenschein*

# INTRODUCTION: Competition and Competition Policy in Market Economies

ALEXIS JACQUEMIN
*Université Catholique de Louvain*

The core problems of the economic organization of society are to determine what commodities shall be produced and in what quantities, how shall they be produced and for whom. Different economic systems try to solve these questions differently, and in a free enterprise society, they are determined primarily by competitive markets. Beyond the various models of competitive equilibrium and the idea that perfect competition in some sense achieves efficiency in the maximization of individual satisfactions, is the general presumption, shared today by most industrialized countries, that a competitive market system is the least bad for promoting economic and political freedoms as well as being the best chance of achieving a high standard of living. Eastern European countries themselves are now discovering in full the failure of social planning.

But it is also recognized that competitive markets do not emerge and survive automatically, that in our decentralized systems imperfect competition, not perfect competition, is the prevailing mode and that there are important market failures.

The question then arises: which governmental policies, if any, could improve the efficiency of the system, at least cost? (for a general analysis of this question, see the contributions in Samuels [1989]).

An important part of the answer is that competition policy is playing a privileged role in many countries, given the fact that it is the most impersonal and least discriminatory means of social control of an economy. As the Economic Council of Canada (1969) has said: 'where competition is such as to promote the efficient use of manpower, capital and natural resources, it obviates or lessens the need for other forms of control such as more or less detailed public regulation or public ownership of industry.'

Still the role and the enforcement of competition policy vary from country to country, and, in the same country, from decade to decade, reflecting changes in economic and social conditions as well as reversals in academic thought. A good example is the treatment of increased market concentration, whether through mergers or other forms of agreement. In the 1960s, European policy sought to exploit the positive link between size and competitiveness which was widely believed to exist, and this meant a green light for mergers which transformed the corporate economy in Europe. But during the economic crisis of the '70s, European super-firms were too rigid and too slow in initiating and responding to change. It is only today that there exists a system of *ex-ante* control of Community-wide mergers at European Community level. Economic thought on the subject of the effects of increased concentration has changed even more than government policy. At times it is, or has been, fashionable to emphasize the static consequences of concentration, such as deviations from marginal-cost pricing, which reduce welfare. However, at other times, the alleged dynamic effects, such as the provision of new products, production processes, and organizational techniques, have been said to be much more important and to be fostered by a more concentrated industrial structure (for an analysis, see Jacquemin and Slade [1989]). Differences in attitude and legislation about cartels and monopolies are also evident from country to country.

Several factors play a role. For example, contrary to large countries, small developed open economies having small effects, if any, on the prices of traded goods maximize their real income level by a policy of free trade; and in this situation antitrust policy is not usually at work given that international trade provides a very effective source of discipline on market performance. In developing countries, on the other hand, the infant industry argument is often used for justifying protection against imports as well as the adoption of various forms of regulation restricting competition at home.

More profoundly, there are differences in attitudes toward economic power (private or public), freedom of contract and trade, efficiency and equity, that are grounded in dissimilarities in political, cultural and moral history, what Edwards [1967] calls the cultural inheritance. For example, for a long time in Europe, though competition has been considered desirable, it has not been accepted as an automatic selective device by which the fittest survive. Instead emphasis was put on the moral obligation in economic affairs: competition has to be fair. The predominant American view, on the contrary, has relied upon the interplay of selfish motives in competitive markets that Adam Smith's invisible hand is able to convert into public virtues.

It is therefore not surprising that the *goals* pursued by competition policy and their relative weight can vary from country to country and over time. These goals can be grouped into three main types.

One eventual goal is the *diffusion of economic power*, the protection of individual freedom and individual rights. Monopolies and cartels can then be seen as a radical, departure from such individualism. It is in the light of these 'non-economic values' that Mestmacker [1980] has characterized the attitude adopted by the German authorities with respect to cartels before World War II. 'The Nazis', he wrote, 'had shown how to transform a highly concentrated and cartelized economy into a central planning system ... Boycotts and collective discrimination were applied against outsiders in order to discipline them in the public interest. If the more traditional measures of economic coercion proved insufficient for the purpose, even the formal transformation of private cartels into compulsory cartels was provided for after 1933' (p. 388). Mestmacker adds that acceptance of cartels was not limited to conservatives who cherished them as safeguards against the anarchy of free competition. Marxists also looked upon cartels and concentration as forerunners of rational socialist planning. He quotes Hilferding, who interpreted this development as tending towards 'a universal cartel, that is a rationally regulated society'! This aspect, which was originally basic to antitrust legislation, still occupies an important place, although perhaps more at the level of public opinion that at policy level.

A second eventual goal of competition policy may be to *protect the economic freedom of market competitors*. Here the protection of competitors takes precedence over the defence of the competitive process as such. Attention is directed towards abusive practices such as coercion, discrimination, refusal to sell, boycotts, and cartels through which powerful firms might endanger the existence of weaker competitors. This type of approach is particularly in evidence in the national laws of 'unfair competition'. According to the Paris Convention of 1883, unfair competition is 'any act of competition contrary to honest practices in industrial or commercial matters'. The corresponding laws are intended to ensure that the competitors compete in a fair way, and carry out their social functions according to an ethical code of conduct. The standard of business ethics plays an important role in developing such a code of honest trade practices, but it is ultimately determined by the common sense of the courts.

A distinction can then usually be made between competition policy concerning efficiency and market injury, and competition law concerning unfair conduct and private injury to one or a few firms. In most instances, like the law on boycotts, fairness and efficiency require the same outcome, but there are situations where a conflict could arise.

The third type of goal of competition policy is dear to the hearts of economists. Competition policy is one of the main instruments to *assure consumer welfare through both allocative and productive efficiency*. The neatest affirmation of a purely efficiency-directed competition policy has been made by Bork (1976). According to his view, antitrust law must challenge inefficient conduct. A necessary (but not sufficient) attribute of inefficiency is a restriction of output beyond levels which would prevail under competitive conditions. Conduct not so identified must be presumed to enhance efficiency, and should not be the subject of legal sanction.

However, in recent years new research in industrial organization has shown that simple formulas for efficiency appear to be deceptive and misleading (see Jacquemin [1987]).

Once the neoclassical paradigm is abandoned, there is no longer the kind of general theorem about the Pareto optimality of the methods of strategic competition that we have for perfect competition. The results are at best ambiguous. Furthermore, with the various types of non-price competition, consumer welfare becomes more multidimensional and includes aspects such as the quality of the product, the speed and security of the supply and so on. Most of these aspects are not measurable. Value judgements are necessary to determine, for example, whether allocating a greater amount of resources to activities which result in technological change or product variation than would be allocated under a more 'classical' form of competition contributes enough to consumer welfare to outweigh the possible losses resulting from static inefficiencies. On the whole, a precise definition of the 'efficiency' criterion is more apparent than real and most of the time requires a delicate appreciation of complex trade-offs.

Not only are there different possible goals for competition policy, but the *means and criteria* used, both for diagnosis and remedy, also differ from country to country and over time.

Some legislation puts the emphasis on the characteristics of *market structure*: market share and the degree of concentration, growth or stagnation of the market, level of barriers to entry, and so on. The policy solution chosen could also be a structural measure such as breaking heavily concentrated markets, disentanglement of company structures, granting or supervision of compulsory licences.

Other types of legislation focus more on *behavioural aspects*: In this case it is some type of behaviour by a firm, e.g. the tying of sales and/or other types of discrimination, which constitutes a restriction of competition, and the policy action takes the form of forbidding the firm to continue behaving in such ways.

In a third type of legislative system the main criterion is *performance*. Intervention here depends on the results achieved by the competing or colluding firms, such as the level of profit margins or technico-economic progress, and sanctions will aim to control some or all of the performance characteristics.

In reality legislative systems are mixed and will emphasize different criteria at different times. Thus US policy has frequently used structural tests, although behavioural criteria have also been applied. For example, US policy towards mergers and dominant firms between 1919 and 1951 were orientated towards the abuse principle, but since the early 1950s emphasis has increasingly shifted towards structural criteria, and, very recently, efficiency considerations. In the United Kingdom several Monopolies Commission reports have shown concern for performance aspects, while in continental Europe a mixture of conduct and performance criteria has long prevailed. This is especially the case in France, with the Conseil de la Concurrence, while the German system is closer to the American one.

One essential aspect is the role of the so-called 'efficiency defense' or, more broadly, the 'public interest' criterion. In contrast to the US tradition, most legislation considers that some behaviour restricting competition in a non-minor way may be exempted because there are sufficient beneficial effects. The usual cases correspond to situations where there is a presumption of market failure such as in high-technology industries. The more weight that is given to such broad criterion, the larger is the discretionary power of the antitrust authorities, and the higher is the danger of confusion between competition policy and industrial policy.

The contributions to this volume represent a major effort to present in one book the various aspects of competition policy just mentioned. It is the first one combining an analysis of behaviour that could reduce competition, in the light of the recent developments in the field of industrial organization, and a presentation of national and international experience in the elaboration and implementation of competition policy. In doing so, the authors illustrate the close connection between the market process and institutions. On the one hand, economic forces are to a large extent responsible for the types of institutions and regulations that arise and for the ultimate shape they adopt; on the other hand, institutions and policies, once formed, influence the operation of the economic process. The case of the European Community, which is progressively dismantling the barriers between its national markets and being concurrently submitted to a new and evolving common competition policy, is an especially clear illustration of the tight link between law and economics.

Through its comparative approach and its interdisciplinary dimension, it is our hope that this book will provide fruitful insights into the complexities of competition in our decentralized economies and help us avoid too parochial views about the roles of competition policy.

## BIBLIOGRAPHY

Bork, R. (1976), 'The goals of antitrust policy,' *American Economic Review*, May.
Economic council of Canada (1969), *Interim Report on Competition Policy*, Ottawa.
Edwards, C. (1967), *Control of Cartels and Monopolies, An International Comparison*, Oceana Publications, New York.
Jacquemin, A. (1987), *The New Industrial Organization*, MIT Press, Boston.
Jacquemin, A. and Slade, M. (1989), 'Cartels, collusion and horizontal mergers, in R. Schmalensee and R. Willig, (eds), *Handbook of Industrial Organization*, North-Holland.
Mestmacker, E. (1980), 'Competition policy and antitrust, some comparative observations, *Zeitschrift für die gesamte staatswissenschaft*, Hefts, September.
Samuels, W. (ed) (1989), *Fundamentals of the Economic Role of Government*, Greenwood Press, New York.

# Economic Foundations of Competition Policy

JANUSZ A. ORDOVER[1]

*New York University*

**INTRODUCTION**

Perfect competition and perfect contestability[2] are but ideals that real markets can rarely if ever attain. Actual markets deviate from these two benchmarks both in their structure and in the conduct of the firms that populate them. Real markets fail. The challenge facing policymakers is to determine what to do about it, if anything, and effectively to implement the desired corrective measures. I do not survey here the gamut of policies that are being used in market economies to affect static and intertemporal allocation of productive resources.[3] Neither

---

[1] Professor of Economics, New York University; Visiting Professor of Economics, Yale School of Organization and Management, and Lecturer in Law, Yale University Law School. I would like to thank Professor Eleanor Fox for most penetrating and helpful comments on the earlier draft. I also received useful comments from other authors in this volume. Alannah Orrison and Ken Rogoza provided excellent research and editorial assistance. Finally, the C. V. Starr Center for Applied Economics at New York University offered generous financial assistance.

[2] Firms operating in a perfectly contestable market earn only a normal rate of return on their assets because any higher rate of return would induce instantaneous entry. For a more precise definition of contestability and conditions that ensure it, see Baumol, Panzar, and Willig [1982, 1986].

[3] Static resource allocation refers to the efficiency with which currently available resources—labor and human capital, physical assets, land, etc.—are deployed to satisfy current private and social wants. Intertemporal resource allocation refers to the efficiency with which currently available resources are allocated between present and future private and social consumption. It is entirely possible for a market economy to exhibit static efficiency along with intertemporal inefficiency. This will occur, for example, when the economy underallocates resources to capital investment, to research and development (R&D), or to the creation of human capital.

do I provide explicit guidelines that would enable decision-makers to select those policies (which, *per force*, include doing nothing) that would be most effective in correcting a perceived market failure.[4]

In this introductory section, I confine my discussion to *competition policy*—or, even more narrowly, to what in the US is referred to as *antitrust policy*—as one among many possible remedial policies. My task is to analyze the economic underpinnings of competition/antitrust policy and to explore its limits and limitations.

Plainly, an account of competition policy emphasizing economic foundations to the exclusion of almost everything else is rather inadequate, possibly misleading, and surely incomplete. Insofar as competition policy based exclusively on economic foundations would be concerned solely with allocative efficiency, as measured by a sum of consumer and producer surpluses,[5] then it is clear that competition policies in the countries discussed in this volume (and those implemented by the EC under the Treaty of Rome) do not build on economic foundations alone. Even in the US, where the efficiency-based approach to antitrust policy and enforcement made spectacular gains during the Reagan Presidency (1981-88), diverse social and political factors remained as viable concerns of antitrust policy.[6] In the EC political as well as economic goals of community-wide market integration continue to affect the formulation and implementation of competition policy.[7]

In short, then, competition policies do not invariably serve the objectives of economic efficiency. This does not mean, however, that there is no purpose to my methodological excursion into the realm of the economic foundations of competition policy. First, the subject of competition policy is firms' conduct. Since conduct affects perfor-

---

[4] For example, such a guideline would assist policy makers in deciding whether a particular firm (industry) should be privatized; whether it should be regulated, or subject merely to antitrust scrutiny, or perhaps even (partly) exempt from such scrutiny altogether.

[5] These terms are defined below.

[6] See the essay by William Comanor in this volume. An article by Lande [1988] provides a discussion of the development of the efficiency approach in the work of Bork [1966, 1978]. It also contains many references to the writings of those American scholars who always rejected efficiency as the only legitimate goal of antitrust. Fox [1986, 1988] and Fox and Sullivan [1987] are especially instructive.

[7] See the chapter by Ken George and Alexis Jacquemin in this volume and Hawk [1986].

mance and allocative efficiency is a measure of performance, it is only appropriate to use efficiency criteria—among others—to ascertain the effects of policies that influence conduct. Second, the opponents of the efficiency-based antitrust agree that efficiency considerations cannot be completely shunted aside in designing policy (see, e.g., Fox [1988]). Third, the proponents of the efficiency approach argue that non-economic (meaning non-efficiency) considerations can be accommodated within the approach as side constraints.[8] I doubt, however, whether this can be easily done. 'Non-economic' values tend to be 'soft,' and thus not readily quantifiable, which makes explicit modelling of the efficiency/non-efficiency trade-off very difficult. Fourth, there is still a fair amount of confusion as to the actual content of the economic, i.e., efficiency-based, approach to competition policy.[9] This section will lay out the basic components of the approach, will argue that the approach need not become ossified in a static vision of a marketplace, and will suggest that a more expansive reading of the economic approach can resolve some of the tensions between the proponents and critics of the approach.

## 2. THE NORMATIVE ECONOMICS OF COMPETITION AND MONOPOLY

Why does one want to have competition policy in the first place? What objectives should it serve? In the introduction, I have already emphasized the fact that competition policies do not invariably serve effi-

---

[8] Technically, the efficiency criterion establishes $W = CS(a) + \pi(a)$ as the appropriate welfare measure for policy assessment, where CS is consumer surplus; $\pi$ is producer surplus, or profit; and a is the vector of policy actions. Giving more weight to consumers' welfare requires that the criterion be rewritten as $W' = [\lambda \cdot CS] + [(1 - \lambda)\pi]$, $1 \geq \lambda > \frac{1}{2}$. The limiting case of $\lambda = 1$ implies that competition policy should aim at protecting consumer surplus (or welfare); see Lande [1988].

[9] For example, Bork's [1966] discussion fails to clarify the distinctions among consumer surplus, producer surplus, and net aggregate economic welfare as appropriate objects of economic policy. At the other extreme, Lande focuses entirely on the price that consumers pay, without mentioning quality or the fact that, in the United States, a significant share of supracompetitive profits accrues to consumers through their participation in pension plans and to workers employed in industries that generate supracompetitive profits. See, e.g., Salinger [1984] on the latter point. Bork's discussion of various surpluses is examined in Lande [1988].

ciency objectives. In this subsection, I abstract from various political and social considerations that can be legitimately pursued within competition policy, and instead give an economic account of the plausible goals of competition policy. These are succinctly stated by Vickers and Hay [1987, p. 2]. According to their definition, competition policy ought to '... promote and maintain a process of effective competition so as to achieve a more efficient allocation of resources.'[10]

There are two key elements to this statement of purpose. The first element is the characterization of competition as a process. From this perspective competition is not perceived as a set of abstractions that describe the general equilibrium conditions in a mathematical model of a hypothetical economy. Rather, competition is seen as a continuous vying for customers—either in a quest for a share of a lucrative market or in an effort to prevent a loss in market share. Spurred on by the profit motive, active participants in the process have strong incentives to offer products and services that match consumers' preferences. However, firms also have the incentive and, at times, the ability to derail competition, to weaken it, to ease the constraint that it imposes on their ability to earn above-normal profit. It goes without saying that this vying for customers does not take place among existing rivals only. Incumbents are constrained in their actions by prospective rivals: i.e., firms that could be attracted into the market by the prospects of adequate returns on their investments.

The other element in Vickers' and Hay's 'statement of purpose' is the maintained focus on efficient resource allocation. There are three distinct aspects to resource allocation. The first, and most familiar, is static efficiency. The second is dynamic efficiency. And the third is the internal efficiency of firms. In most cases, the competitive process conduces to both static and internal efficiency. The relationship

---

[10] In *Olympia Equip. Leasing Co. v. Western Union Tel. Co.*, 797 F. 2d 370, 375 (7th Cir. 1986), Judge Posner wrote that '[T]he emphasis of antitrust policy [has] shifted from the protection of competition as a process of rivalry to the protection of competition as a means of promoting economic efficiency . . .' The Posnerian dichotomy is overreaching. Indeed, the process of rivalry generally conduces to static economic efficiency. Moreover, given the sources of antitrust policy in the United States, why should its formulation and implementation be governed by 'end state' principles, such as efficient allocation of resources? Professor Fox suggests that there may be an independent value in preserving competition as a 'process.' See, e.g., Nozick [1974, ch. 7] for a discussion of end-state versus process-based moral philosophies.

between competition and dynamic efficiency is much more complex, as was already noted by Schumpeter.[11] However, even when it comes to static efficiency, there are potentially significant tensions between competition and efficient resource allocation.

The core problem of competition policy from the vantage point of economics is, to my mind, the appropriate identification of salient exceptions and the design of workable, simple rules that would enable the fact-finder and decision-maker to strike a reasonable balance between the twin goals of encouraging and maintaining competition and ensuring efficient allocation of scarce economic resources.[12]

We can illustrate this tension with an example of a tie-in, whereby a seller requires that a consumer buy products A and B1 from it. A tie-in apparently undermines the competitive process because it prevents a seller of B2, which is an excellent substitute for B1, from competing for those who buy product A.[13] A closer inspection reveals that in many settings that tension dissolves. This happens if (i) without a tie, A could not be offered at all; if (ii) sellers of B2 could offer a competing deal by joining with C; or if (iii) a tie pertains to a small portion of the customer base and B2 is not disadvantaged in its other sales by not being able to sell to A's customers. Obviously, one could also construct other scenarios where tie-ins do harm the competitive process. For example, Whinston [1989] demonstrates that tying can change the market structure for the tied good, increase the profit of the tying firm and possibly harm social welfare.

I now briefly explicate the traditional normative approach to competition, monopoly, and economic efficiency. Then I turn to the dynamics of competition.

---

[11] See, e.g., Reinganum [1989], Baldwin and Scott [1987], and Ordover and Baumol [1988] for summary discussions of the potential trade-off.

[12] It is often stated that the goal of US antitrust law is to protect competition and not competitors. I agree. However, I would suggest that in the absence of viable actual and/or potential competitors, competition is not likely to be very effective. See note 10, above.

[13] A tie-in also restricts the freedom of choice of those who buy A and prefer B2 to B1. Such a restriction could be regarded as an undersirable feature.

## 2.1. Static deadweight cost of monopoly[14]

In a perfectly competitive market for a homogeneous good, the long-run equilibrium price equals marginal cost, ensuring that every firm active in that market earns only a normal rate of return on its assets (after allowing for the Ricardian rents).[15] In Figure 1, the competition price is p*; competitive output is q*. Assume now that a putative monopolist restricts output to $q^m$ and raises price to $p^m$. The deadweight cost ($DW^s$) of monopoly conduct can then be estimated by the area of the shaded triangle, ABC.[16] This triangle measures the portion of the aggregate loss in consumer surplus that does not accrue to the monopolist in the form of higher profit. The rectangle $ABp^*p^m$ reflects that portion of the lost surplus that is captured by the monopolist. Simple calculation shows that the area of the deadweight loss triangle ABC is given by:

$$DW^S = (1/2) \cdot \epsilon \cdot [(p^m - MC)/p^m]^2 \cdot (p^m q^m),$$

where $\epsilon$ is the elasticity of demand and MC is marginal cost, which is equal to p*. In the event that $p^m$ results from a profit-maximizing decision of the putative monopolist, the above expression can be rewritten as

$$DW^S = [(1/2) \cdot p^m q^m] \cdot L = [(1/2) \cdot p^m q^m]/\epsilon, \qquad (2.1)$$

where L is the Lerner Index of market power,[17] i.e.,

$$L = (p^m - MC)/p^m = 1/\epsilon. \qquad (2.2)$$

---

[14] My exposition here borrows from Schmalensee [1982].

[15] The discussion here assumes that perfect competition is a feasible alternative to imperfect competition. Alternatively, we could assume that p* is the price that would obtain if the market were perfectly contestable.

[16] An additional welfare loss would occur if the marginal cost curve were upward-sloping. But how realistic is such a case in an industry that is highly concentrated, i.e. in which market power would, in fact be possibly exercised? A reasonable answer seems to be that it is not very likely. This is due to the fact that concentrated industries are usually characterized by scale economies, and in such industries pure producer rents—measured by the difference between marginal and average cost (MC-AC)—do not arise. Also, I abstract from the possibility that there would be pure efficiency losses, if the 'monopolist' were carved up into competing enterprises. Hence, p* would best be regarded as the price that would obtain under perfect contestability of the monopolist's market.

[17] I will discuss the Lerner Index at greater length below.

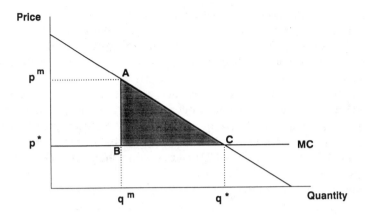

FIGURE 1

As equation (2.1) indicates, the deadweight loss is related both to the size of the markup, L, that the monopolist can realize and to the size of its market, as measured by revenues realized at the profit-maximizing price.[18] Indeed, a monopolist with a high Lerner Index, but small revenues, imposes only a small deadweight loss on society. Consequently, scarce prosecution resources perhaps ought to be devoted to investigating conduct of firms that could impose large deadweight loss on the society.[19]

For several reasons, formula (2.1) can overestimate or underestimate the social cost of the exercise of monopoly (market) power, as calculated by the sum of lost consumer surplus and gained producer surplus (profit).[20] I will elaborate on these reasons below. Here I point

---

[18] Similar formulas can be derived for more complicated market scenarios. In all possible market scenarios, the deadweight loss imposed by a firms is the 'triangle' under the firm's residual demand curve. See Schmalensee [1982]. Note also Ordover, Sykes, and Willig [1982].

[19] See Landes and Posner [1981] and Schmalensee [1982]. However, if the probability of prevailing is positively related to the size of L, prosecutors are likely to select such firms for enforcement. Similarly, conducts that are blatantly anticompetitive will most likely attract prosecution resources.

[20] It is clear that whenever a firm faces an imperfectly elastic demand curve for its product, it will be able to exercise some *market power*. Firms with significant degrees of market power are often said to possess *monopoly power*.

out that (2.1) is agnostic about distributional issues; i.e., it reports a dollar loss in consumer surplus on par with a dollar gain in profits (or monopoly rents). But what if competition law assesses conduct purely from the standpoint of its effect on consumer welfare, rather than on aggregate welfare?

Firstly, maximization of consumer welfare need not in general conflict with the goal maximizing aggregate welfare, provided that policy-makers are mindful of the side constraint on firms' profits. This is so because in a variety of situations a policy that increases consumer surplus by one dollar (or ECU) causes a less than equivalent fall in rents. But this is not always true. For example, Shapiro [1986] has demonstrated that allowing oligopolists to exchange certain types of strategic information raises expected profits by more than it lowers consumers' welfare. The query is how often such unusual trade-offs actually arise.

An additional point is this: if profits are to be disregarded in a welfare measure, why not also give a lesser welfare weight to consumer surplus accruing to richer consumers? Then antitrust policy can be deployed as a redistributive tool for shifting income from richer to poorer consumers. Thus, assume that poorer consumers are more concerned with price relative to quality than are wealthier consumers. Then, a policy that impedes quality-increasing vertical constraints and keeps price low has the described redistributive impact.

It is difficult to establish that on the whole an average consumer who is injured (directly or indirectly) by exercise of market power owns a below-average share of national wealth (see, however, a pioneering study of the issue by Comanor and Smiley [1975]). Indeed, consumers-cum-workers own a substantial share of national wealth through pension plans and, as shown by Salinger [1989], they also share in monopoly rents through higher wages.

All these reasons suggest that an aggregate welfare-oriented standard provides a sound foundation for efficiency-driven competition policy.

Let me now consider how well formula (2.1) reflects the aggregate welfare loss from the exercise of market power.

*2.1.1. Assumption of a single price*
First of all, the formula is based on the assumption that the seller charges a single price for its product. This is too restrictive. Sophis-

ticated pricing techniques enable firms to capture a greater portion of the available consumers' willingness to pay, and thus increase profits.[21] For example, a firm can price discriminate.[22] It can sort buyers into different groups by confronting them with a menu of two-part tariffs.[23] It can use tie-ins,[24] commodity bundling,[25] seasonal discounts, and so on. These pricing strategies are profitable to firms using it. Their effects on aggregate welfare are not always beneficial (see Phlips [1983], Varian [1989], and Tirole [1988, ch. 3]) Sophisticated pricing is likely to increase aggregate welfare if it induces output expansion, or allows the seller to recoup fixed costs that otherwise would not be recovered. But such pricing techniques can harm some buyers and benefit others, as compared to the level of surplus that would be realized under a single price.

These pricing techniques can increase profit without affecting the intensity of rivalry, i.e., the competitive process.[26] These arise from better extraction of the available consumer surplus. The firm deploying them more effectively exploits the market power inherent in the assets it owns. Should efficiency-based competition policy be concerned with such forms of exploitation of market power? Unequivocal answers are not possible. Two points are worth making, however. First, if sophisticated pricing is deployed strategically to exclude rivals or induce exit—to lessen the competitive constraint facing the firm—it surely merits antitrust scrutiny. Second, in some markets, sophisticated pricing on the whole conduces to economic efficiency. I have in mind, in particular, markets for information. It is well known that private incentives to produce and disseminate information are frequently lower than the social ones because of the spillover ad appropriability problems. Consequently, when information is the commodity transacted (e.g., in patent licensing) sophisticated pricing and contractual

---

[21] A perfectly discriminating monopolist can capture all of consumer surplus and sell $q^*$ units of output. This maximizes profits and wipes out the deadweight loss.

[22] For example, as when students and faculty members pay different prices for scholarly journals.

[23] For example, as when a golf club has different membership plans.

[24] For example, as when an owner of a fried chicken franchise requires that the franchisees purchase paper buckets, in which chicken is taken out from the store, from the franchiser.

[25] For example, as when a manufacturer of central processing units (CPUs) bundles certain peripheral devices with the CPU.

[26] Various forms of vertical restraints have similar consequences.

restraints can greatly improve appropriability and profitability and cure some of the market failure.[27]

### 2.1.2. Assumption of fixed quality

The deadweight loss of monopoly in equation (2.1) is calculated for a given quality of the product offered by the seller. It shows that, whatever the quality, the product will be overpriced by a firm with market power. However, the quality offered by such a firm may also deviate from the socially optimal level, thereby imposing another welfare loss. There is no presumption, of course, that imperfect competition leads to underprovision of quality.[28]

Quality decisions attract antitrust scrutiny most often when a firm resorts to vertical restraints to increase the level of quality that otherwise would be provided. For example, assume that the 'quality' of the manufacturer's product depends on the level of services provided by the dealers. To induce its dealers to provide a desired level of services, the manufacturer may have to resort to vertical restraints, such as territorial restrictions or retail price maintenance. The resulting equilibrium price and quality package need not generate a higher level of aggregate welfare as compared to the level that could be attained in the absence of such restraints, but it may. Competition policy usually does not interfere with manufacturer's direct decisions regarding quality.[29] One might argue, therefore, that indirect decisions should be accorded similar treatment. The difference between direct and indirect decision-making is that (i) vertical restraints do in fact restrain downstream competition among distributors of the product in question; and, (ii), unlike direct decisions, indirect decisions are transparent and easily attacked. Reason (i) might compel one to an overly narrow interpretation of the competitive process. And reason (ii) is not compelling at all, in my opinion.[30]

---

[27] In addition, various other transactional considerations, besides appropriability, may mandate the use of sophisticated pricing in information markets. For more on this in an antitrust context, see Ordover and Baumol [1988], which provides a lengthy bibliography as well.

[28] See Scherer [1980] and Tirole [1988, ch. 3].

[29] Thus, a light-bulb manufacturer is free to determine the durability of its light-bulbs. See Pynchon [1973] for an interesting discussion of light-bulb durability problems.

[30] To say that competition policy should favor lower price to more service, or the advantage of price competition among dealers to more service, creates a distinction between price and quality which is without basis in economics. Perhaps the hidden reason here is redistributive on the presupposition that low income consumers value marginal quality improvements less than high income consumers?

## 2.1.3. Marginal cost pricing

Formula (2.1) overestimates the cost of monopoly when pricing at marginal cost would be unprofitable. See Baumol, Panzar, and Willig [1982] for a complete exposition of the feasibility of marginal cost pricing.

## 2.1.4. The status of monopoly rents

In an influential article on the social cost of monopoly, Posner [1975] argued that the profit rectangle (denoted by $ABp^*p^m$ in the figure above) is also the cost of monopoly. His conclusion requires two assumptions: (i) there is free entry into the competition process of becoming a monopolist, so that the monopoly rent is fully dissipated in equilibrium; and, (ii) all the rent—dissipating expenditures made by the seekers of market power are socially wasteful. Realistically, neither one of those assumptions is likely to hold. For example, some firms may be more effective (and efficient) at capturing monopoly rents, which means that only a portion of the rectangle will be dissipated. In addition, for strategic and other reasons, those firms that are currently earning the monopoly rent may be in a better position to earn it the next period and thus will not have to dissipate all the monopoly profit in an effort to maintain their dominant positions.[31]

It is clear also that rent-dissipating expenditures need not be socially wasteful. A few examples will clarify this point. Firms invest in R&D in the expectation of *ex post* monopoly profit. If entry into R&D competition is free, *ex ante* profit will be driven to zero. This could mean that too many firms will be vying for the monopoly profit from a successful innovation.[32] For another example, in a market totally unprotected by an entry barrier, the monopolistic incumbent dissipates all of the rent by expanding output to the level that yields no supracompetitive rate of return.[33] Finally, in a Chamberlinian version of competition with differentiated products, introduction of new varieties continues until no variety makes more than a normal rate of return. Koenker and Perry [1981] demonstrated that unrestrained entry—i.e., the competitive process—can produce too many varieties of the substitute

---

[31] See, e.g., Fisher [1985] for reasons that (i) need not hold. Some additional arguments can be found in Shepherd [1988].

[32] See Reinganum [1989] for a discussion of rent dissipation in R&D competition.

[33] This outcome obtains in perfectly contestable markets. For more, see Fudenberg and Tirole [1987].

products. Still, some variety is desirable in equilibrium, so only a portion of the aggregate expenditure on quality is socially wasteful.

In sum, the aggregate reduction in consumer surplus, given by the trapezoid p*p$^m$AC, in Figure 1, most likely overestimates the social cost of monopoly. On the other hand, the deadweight loss is not always well approximated by the usual triangle. Moreover, the shift from competition to 'monopoly' cannot normally be captured by the simple diagram of Figure 1. A shift in industry structure also implies a change in industry costs and in the qualities and varieties of products offered on the market. Consequently, such a shift affects the relevant cost and demand curves. This should be borne in mind when considering the possible scope and limits on structural remedies for market power.

## 2.2. Dynamics of the deadweight cost of monopoly

Figure 1, and the associated formula in equation (2.1), is a mere snapshot of the market. In any given market, the ability of a firm (or firms) to repress output below 'the' competitive level and to earn supracompetitive returns is likely to vary over time. Unless the incumbents have a permanent ability to exclude competition completely, the process of competitive entry will tend to whittle down the incumbents' market power, and excess profit, below the short-run level, if not down to zero.[34] Again, staying with the market for a homogeneous product, Schmalensee [1982, p. 1794] demonstrates that the total deadweight loss, TDW, is given by the following formula:

$$\text{TDW} = (1/r) \cdot [(r \cdot \text{DW}^S) + (\gamma \cdot \text{DW}^L)] \cdot (r + \gamma)^{-1}. \tag{2.3}$$

In the formula, r is the interest rate used to capitalize future welfare losses, and $\gamma$ is '... the annual fractional reduction in the gap between short-run, [DW$^S$], and long-run [DW$^L$], [deadweight loss]—in effect, the rate of decay of market power...' held by the incumbent dominant firm, or a group of firms.

The discussion of the dynamics of dominance and monopoly can be

---

[34] The reader should understand the short-run deadweight loss as an instantaneous loss per unit of time (day, week, month), while the long-run deadweight loss is the steady state level. A simple model in which firm's market power declines smoothly to zero has been developed by Gaskins [1971]. I will comment on it later on.

usefully organized using Schmalensee's formula in equation (2.3).[35] It draws attention to three critical concerns of competition policy: (i) the strategies for acquisition of market power—i.e., the strategies firms employ to place themselves in a position to inflict a $DW^S$ on the society; (ii) the rate of decay of market power and the strategies firms employ to slow it down; and (iii) the long-run level of market power $DW^L$.

Another way to think about formula (2.3) is by making a distinction between tactical and strategic conduct variables at the disposal of a firm. Tactical variables, such as pricing policy or product quality, determine the magnitude of $DW^S$, which can be viewed as the amount of market power that can be exercised given the current state of competition. Strategic variables, such as investment in R&D, investment in plant and equipment, and investment in marketing (including reputation), have a long-lasting impact on the market and are not readily reversible—they have a precommitment value. The choice of strategic variables affects the evolution of the market and hence the amount of surplus that will be available for extraction in the future.

In terms of our formula, we could think of $DW_t^S$ at time t as being determined by a vector of tactical variables $v_t$ for any given history of strategic variables $<s_0,...,s_t>$.[36] That is, we have $DW_t^S(v_t, <s>)$. Then, the choice of $s_t$ affects $DW_{t+1}^S$, $DW_{t+2}^S$, ..., etc.

The formula indicates that if the erosion of market power is rapid and if market power dissipates in the long run, then competition policy play a secondary role in maintaining rivalry and ensuring efficient allocation of resources.[37] On the other hand, if market power decays slowly and is never fully dissipated, then competition policy can significantly improve the functioning of markets, beyond what could be accomplished through self-correcting market forces. There is some debate as to which of these descriptions better reflects the realities of market dynamics in the preponderance of cases.[38]

---

[35] See Gilbert [1988, 1989], Geroski, Gilbert, and Jacquemin [forthcoming], Mueller [1986], and Tirole [1988], for excellent discussions of the value of incumbency and dynamics of market power. Shepherd [1988] severely criticizes much of the post-70's literature on the value of incumbency from the vantage point of what he characterizes as 'mainstream industrial organization' theory.

[36] The distinction between tactical and strategic variables is not as neat as I make it sound. For example, when price is used for 'predatory' purposes, it acquires the dimension of a strategic variable. See Ordover and Saloner [1989].

[37] If $DW^L = 0$ and $\gamma \to \infty$, then $TDW \to 0$.

[38] Compare Gilbert [1989], whom I follow, and Shepherd [1988].

At one end of the theoretical spectrum, we have two hypotheses about market dynamics and the efficacy of the threat and speed of entry as corrective mechanisms. These hypotheses are captured by the model of perfectly contestable markets (PCM)[39] and by the 'Chicago School' model of efficiency-based differences in profitability and market share.[40]

Fundamental to contestability is the assumption that a potential entrant takes the incumbent's current, i.e. pre-entry, prices as the correct predictor of the incumbent's post-entry prices.[41] Given these expectations, for entry to occur the potential entrant must be able to earn a positive profit by charging prices that are (slightly) lower than the incumbent's. The entrant's prices must generate enough surplus to defray the sunk costs of entry. Sunk costs are that portion of entry investment that would be lost if the entrant had to exit the market before its investment was fully depreciated. Clearly, the entrant cannot expect that the incumbent will stick to its pre-entry prices forever. However, for profitability calculations it suffices that the incumbent sticks to its prices long enough for the entrant to recoup sunk costs. Roughly speaking, the smaller the sunk entry costs, the shorter is the period of time needed for their recovery; hence the smaller is the entry risk. In the limit, when there are no sunk entry costs the market is perfectly contestable. An entrant can invade a perfectly contestable market for an 'instant' and still earn a profit, if the incumbent's prices are too high. The incumbent realizes that 'too high' prices attract entry and generate losses. Therefore, the incumbent sets prices at the level that deters entry. In a perfectly contestable market—where sunk costs are zero—entry-deterring prices (if such exist) yield only a normal rate of return for the incumbent.[42] In terms of our formula, then, the PCM

---

[39] The theory of contestable markets is summarized by Baumol, Panzar, and Willig [1986] and Baumol and Willig [1986]. Schwartz [1986] and Stiglitz [1988] are good examples of critical assessments of contestability. Shepherd [1984, 1988] provides another critique of contestability. Unfortunately, his discussion is marred by a misunderstanding of the theoretical foundations of contestability.

[40] Stigler [1968, ch. 7] and Demsetz [1973] provide clear statements of the Chicago school position.

[41] In assessing the likelihood of entry, the 1984 Department of Justice Merger Guidelines ask whether entry would occur if prices were to increase by five to ten percent above the current level and *remained* there for two years.

[42] If the monopolist produces only one product, it will set price equal to average cost. See Baumol, Panzar, and Willig [1982] for the analysis of pricing by a multiproduct incumbent.

hypothesis implies that $\gamma = \infty$ and $DW^L = 0$, giving us $TDW = 0$.

The core of the 'Chicago School'/Demsetz paradigm is the notion that differential efficiencies across firms explain differences in rates of returns among firms, differences in market shares, as well as the empirically documented correlation between an industry's concentration and its average rate of return. The theoretical explanation for these phenomena is that firms with lower production costs, superior products, more effective management structure, or superior marketing and distribution networks, gain market advantage over less efficient rivals. These firms become larger than their less efficient rivals and earn rents for their differential efficiency.

From this perspective, a measure of market power that is based on price-cost margins is potentially quite misleading. But there is more to the Chicago position. Its focus of efficiency rents is hardly controversial. After all, the search for such rents is the driver behind the competitive process. What is controversial is the notion that supracompetitive profits stemming from collusive or exclusionary conduct are quickly dissipated by market forces; the only profits that remain in the long run stem from differential efficiencies among firms.[43] If this is so then market forces, rather than antitrust, can be effectively relied upon to mitigate and dispel whatever transitory market power there is.

The Bain-Sylos-Modigliani (BSM) limit pricing model provides an alternative version of market dynamics. In that model, once the firm obtains market power it can protect it definitely, barring exogenous market shocks.[44] Consequently, $DW^S = DW^L$ and $TDW = DW/r$; i.e.,

---

[43] In an interesting recent paper, Thomas [1989] notes that the exponents of this view have been less than clear on how quickly market forces dissipate efficiency rents. Clearly, although this is not Thomas' point, if such rents persist over time this may in and of itself signal some form of market failure and call for policy intervention. Thomas analyzes the Kellogg Company's long-run performance in the ready-to-eat breakfast cereal market and concludes that its persistently above average profits, high Tobin q ratio, and market share, can all be explained by its asymmetric efficiency *vis-a-vis* other actual and potential rivals. Analytically, it is not a simple matter to distinguish this finding from another, which is that the persistence of above-average profits could also be attributed to intertemporal linkages in 'market power.'

[44] Schmalensee [1989] offers a compact formulation of the difference between the BSM and Chicago School/Demsetz models, which can be empirically analyzed in a cross-section. Thus, write $r_i = \rho_i + \delta_i H_i$; where $r_i$ is the average rate of return in industry i, $\rho_i$ and $\delta_i$ are industry-specific constants, and $H_i$ is the industry's Herfindahl-Hirschman Index of concentration. Bain's thesis would be that the second term in the equation is small and that $\rho_i$ and $H_i$ are correlated in a cross-section. The Demsetzian alternative would state that $\rho_i$ varies little across industries, while the $\delta_i$ terms are large and positive, leading to a significant correlation between r and H.

TDW is just the present discounted value of the instantaneous deadweight loss. If the BSM model is right about market dynamics then competition policy could play an important role in improving resource allocation and dissipating market power. The BSM model questions the effectiveness of market forces in dissipating market power and implies that anticompetitive conduct the generates market power can have prolonged effects on economic efficiency. It also indicates that firms have a substantial incentive to expend resources on activities that might generate market power. This is in contrast to the Chicago view, wherein such investments are not likely to be profitable given rapid dissipation of market power by the competitive process.

Gilbert [1989] concludes that empirical evidence does not strongly confirm either the PCM or the BSM model of market dynamics. He finds some empirical support for Gaskins' [1971] model of dynamic limit pricing. In that model, the incumbent steadily loses share to the ever expanding fringe and sees its profit margins fall, conceivably to zero. However, Gilbert remarks that the evidence, as adduced by Mueller [1986], for example, is not entirely inconsistent with the 'Chicago School' version of market dynamics,[45] Shepherd [1988] finds, on the basis of similar evidence, some support for the BSM hypothesis. The debate on the speed with which market power dissipates is not resolved. It would be unrealistic to expect that it will be readily resolved, even with the help of sophisticated econometric techniques.

### 2.3. Dynamic efficiency and market power

The formula for the TDW of monopoly starts counting the deadweight loss from the time market power is acquired or exercised. It does not delve into the sources of market power, and does not calculate the value of the resources that have been expended on its acquisition (see section 2.1.4. above). Clearly, how market power is acquired is of great importance. Market power can be acquired through a variety of conducts, that differ in their consequences for static and dynamic efficiency. Thus, at one extreme, market power can be acquired through collusion, which holds no consumer benefit; through merger that reduces the intensity of market rivalry, but which may also produce some cost savings; or through exclusionary practices, such as price and

---

[45] See also Thomas [1989].

non-price predation. Market power can also be obtained through other means, such as investment in R&D. Competition policy should encourage technological competition despite the fact that such competition could lead to (possibly transitory) market power.

Consider a very simple model of a technological race. The participants in the race expends R&D resources in order to secure a patent. Only one firm can be a winner. The winner obtains an air-tight patent that gives it a right to exclude others from the use of the new technology. The winner is thus a lawful monopolist. Let $V^* = E\pi(b^*(N), N)$ be the expected equilibrium profit of a representative firm that is in a race with N other rivals, each spending $b^*$ on R&D. If there is free entry into the patent race then $V^*$ must fall to zero, so that *ex ante* profits are wiped out.[46] Of course, *ex post* profit will accrue to the winner of the race. *Ex post* profit will have to be high enough to induce the requisite number of firms, but no more, to enter the race. Obviously, *ex post* profit is not a meaningful measure of market power. Modify the model a bit, and assume that only some fixed number of firms can compete in the patent race—there are some entry barriers into R&D competition stage. Then the equilibrium value of $V^*$ will be positive. There will be genuine *ex post* market power. The important point is that the source of *ex post* market power in a downstream market is insufficient competition at the R&D stage. For, as I noted, if there were free entry into that stage, all the expected profits would be dissipated through firms' investments in R&D.[47]

The example demonstrates how market power can arise in industries where dynamic competition is intense and where winners of the competitions tend to gain substantial advantage over other rivals. This temporary advantage acts as a stimulus to R&D investment and is recognized by public policies as embodied in the system of intellectual property rights.

Formula (2.3) focuses on the *product-specific* market power held by a firm over time. The additional intertemporal dimension of market power, which is termed *market-specific* market power in Ordover and

---

[46] Technically $N^*$ will be an integer such that $V^*(N^*) > 0 > V^*(N+1)$.

[47] This does not mean, however, that efficiency would be improved if more firms were to enter into the R&D stage. Note that each firm expands $b^*(N)$ on R&D, so that the industry expands $Nb^*(N)$ in the aggregate. This may be too much, given that only one firm can be a winner. For more on the issue of excessive entry into R&D races see Reinganum [1989].

Baumol [1988], could also be handled by the formula, as long as the calculation of deadweight loss reflects the changing nature of products in the pertinent 'market'. Market-specific market power arises when *existing* market power would be a precondition for successful future innovative effort. This would be the case if (i) R&D were difficult to finance with outside funds, so that a firm with market power could use its current excess profit to fund the next generation of products and technology at a lower cost than some other firm that must borrow from outside; (ii) the monopolist could better appropriate social benefits from its R&D investments. A monopolist can better appropriate such benefits because it can engage in a greater variety of pricing strategies than a highly competitive firm and because a monopolist has fewer free-riders to comport with. The Schumpeterian thesis, that market concentration conduces to R&D investment, provocative as it sounds, has not been established in the theoretical and empirical literature.[48] There is some support for it in the data, pointing in part to efficiency-based, intertemporal linkages in market power that explain the persistence of firms' dominant positions over time in changing markets.[49]

To measure the deadweight loss in a fully dynamic model of monopoly, one must carefully calibrate the 'competitive' benchmark. First, because dynamic competition entails fixed costs (on R&D, promotion of new products, etc.), marginal cost pricing may not be feasible. Consequently, deviations of prices from marginal costs need not signal a breakdown of competition. Second, dynamic competition can lead optimally to market concentration, as in the example above in which the winner takes all. In equilibrium price deviates from marginal cost, so that again a comparison of price to marginal cost is an inadequate guide for policy. In fact, such comparisons could be counterproductive if they engender corrections that signficantly reduce the rewards from technological races and thereby undermine the incentives for technological progress. And, finally, in light of what we just concluded, the 'dynamicized' deadweight loss from static market power has to be counterbalanced against the *truly dynamic* benefits from technological progress. The rate of technological progress can be

---

[48] See Baldwin and Scott [1987] for a review of the pertinent literature.
[49] Plainly, IBM's current product offerings are different from those offered only twenty years ago. On the other hand, Heinz ketchup has not changed over the decades!

affected by policies aimed at reducing the static distortions, as discussed at length in Kaplow [1984] and Ordover and Baumol [1988].[50]

From the dynamic perspective, the relevant concern for competition policy should be to facilitate rather than stymie the process of technological competition.

## 3. INDICES OF MARKET POWER

In their important article on the measurement of market power in antitrust cases, Landes and Posner (LP) [1981] wrote that '[t]he standard method of proving market power in antitrust cases involves first defining a relevant market in which to compute the defendant's market share, next computing that share, and then deciding whether it is large enough to support an inference of the required degree of market power' (p. 938). (The LP position is something of a canard because, in law, share is never enough in itself.) They offered an extended critique of the traditional approach and proposed instead an inferential standard based on the Lerner Index. In this subsection, I review the LP proposal, consider the forensic value of market shares in determining market power, and consider the usefulness of some other indices of market power.

### 3.1. The Lerner Index

Equation (2.1) provided a measurement of the deadweight loss from the existence of market power, which was defined as the ability to raise price above a 'competitive' level' and exclude competitors. In that formula, short-run welfare loss was directly related to the Lerner Index. In turn, the Lerner Index, i.e., the price-cost margin, was inversely related to the *market elasticity of demand*, $\epsilon$, in equation (2.2). Thus I abstracted from the question of what is the appropriate elasticity for assessing firm-specific market power, and, instead, defined this elasticity conventionally as the percentage change in quantity demanded of the product in question induced by a (small)

---

[50] See, however, Ordover and Willig [1981] for a proposition that product innovations can be anticompetitive.

percentage change in its price: $\epsilon = -(\Delta q/q)/(\Delta p/p)$, where q is the quantity demanded in the entire market.[51]

Calculation of the pertinent elasticity raises some methodological questions. Conventionally, this elasticity is calculated on the assumption that prices of all other products, be they substitutes or complements, remain unchanged while the price of the relevant product varies. The elasticity in formula (2.1) reflects a hypothetical situation in which (i) the relevant market is effectively 'monopolized';[52] (ii) entry into the market is fully blockaded, even at supracompetitive prices; (iii) changes in the price of the relevant product are not accompanied by changes for other products;[53] (iv) the putative monopolist (or a perfect cartel) exercises market power over only one product—the one for which the Lerner Index is computed;[54] and (v) there are no intertemporal cost or demand linkages. In sum, the methodological issue is, then, one of selecting the correct elasticity for forensic assessment of firm'(s) ability to raise price above marginal cost.

Indeed, all five assumptions are likely to be violated to some degree in actual markets. In each case, therefore, the Lerner Index will have to be modified to reflect the realities of the market in which the dominant firm (or group of firms) operates. In general, the elasticity of the market demand curve will not indicate the firm's ability to increase the price above marginal cost. A better (in some cases, an exact) measure of the ability of a single firm to elevate price is given by the elasticity of residual demand facing the firm. This elasticity measure captures the strategic interactions between the firm and its rivals. In general, residual demand is likely to be more elastic than is market demand. The remaining discussion demonstrates the role of the residual demand curve in the assessment of market power.

---

[51] Here I abstract from the methodological problems of *market definition*. I take up those in subsection 4. Here, I assume that the relevant market has already been defined and the analytic problem is the assessment of market power itself.

[52] This means that either the market consists of a single firm or the price is set by a perfect cartel selling a homogeneous product.

[53] In this case there is a 'marked gap in the chain of substitutes' so that the putative monopolist is not interacting strategically with other products. The concept of the gap was introduced into the literature by Robinson [1933].

[54] This assumption is necessary because if the monopolist controls more than one product, it may raise all the prices above their respective marginal costs. In this case, even in the absence of the strategic interactions ruled out in (iii), the elasticity facing each product might differ from what it would have been had only one price been raised. For more on pricing by a multiproduct monopolist, see Baumol, Panzar, and Willig [1982].

### 3.1.1. Fringe firms

The simples scenario to consider—which is also the cornerstone of the LP approach—is a dominant firm that competes in the market with price-taking fringe firms. The elasticity of supply of the fringe, $\eta_f(p)$, depends on the price. In this case, the elasticity of residual demand is $\epsilon^r = \epsilon + s_f \eta_f$, where $s_f$ is the market share held by the fringe firms. The adjusted Lerner Index is

$$L = (p - MC^d)/p = 1/\epsilon^r = 1/(\epsilon + s_f \eta_f). \qquad (3.1)$$

Obviously, if the elasticity of supply by the fringe firms is very high (at prices close to marginal cost), then the ability of the dominant firm to increase price above the 'competitive' level is severely restricted. Alternatively, the fringe can be very inefficient, and its constraining effect may not kick in until the dominant firm charges a price substantially above marginal cost. In the former case, calculations using only the market elasticity of demand around p = MC would substantially overestimate market power. In the latter case, on the other hand, the error would be most likely insignificant. In either case, caution has to be exercised in the use of the formula based purely on the information about $\epsilon$.[55]

### 3.1.2. Oligopoly

The Lerner Index must be modified also in the case of an oligopolistic market that is not perfectly cartelized. For the sake of concreteness, assume that there are N > 1 oligopolistic firms, and possibly a fringe, that compete in the relevant market. To make the exposition as simple as possible, I assume that the oligopolistic firms play a Cournot game. That is, each oligopolist selects a profit-maximizing level of output taking as given the aggregate output chosen by other oligopolists. In this case, it is easy to show that

$$L_i = (p - MC_i)/p = s_i/(\epsilon + s_f \eta_f), \quad i = 1, \ldots, N \qquad (3.2)$$

where $MC_i$ is the marginal cost of firm i, and $s_i$ is its market share.

In the market scenario considered here, the residual elasticity for firm i is equal to the rightmost fraction in (3.1). We see that the Lerner index for a firm is an increasing function of the firm's market share.

---

[55] It seems that LP are too harsh on the traditional approach here. The traditional approach can easily accommodate the existence of the fringe and its ability to expand in response to price changes.

That is, the smaller is the firm, the more elastic is its residual demand.[56] However, market share cannot be used to assess market power without information about the value of residual elasticity, as I will discuss in subsection 3.2.

We can construct an industry-wide Lerner Index by aggregating the firm-specific Lerner Indices of equation (3.1). One simple aggregation weighs each firm's Lerner Index by its market share.[57] Then, the aggregate index, L*, is given by

$$L^* = \sum_{i=1}^{N} s_i L_i = \sum_{i=1}^{N} s_i^2/(\epsilon + s_f \eta_f). \tag{3.3}$$

L* is the industry-wide average deviation of market price from industry marginal cost.

L* is inversely related to the usual elasticities. It also depends on the sum of squares of market shares of the oligopolistic firms. That sum is the Herfindahl–Hirschman Index (HHI) of market concentration.[58] The Antitrust Division of the US Department of Justice uses the HHI for evaluation of likely anticompetitive consequences of mergers. Formula (3.3) supports this approach because it indicates that the industry-wide degree of market power increases with the degree of concentration.[59]

### 3.1.3. Entry

The Lerner Index formula does not capture threats of potential entry other than in the simple case of smooth entry-cum-expansion by the competitive fringe. The reason is that entry is a binary event: either it occurs, or it does not. And, when it does occur, there could be a large,

---

[56] There is one subtle point regarding this elasticity that must be noted: Conjectures held by Cournot firms are not consistent. Each Cournot firm is acting as if its residual demand were more elastic than it actually is. In equilibrium, when firm i expands output its rivals contract their output, which reduces actual elasticity. Cournot firms assume, however, that the rivals stick to their output. See Baker [1987] and Bresnahan [1989] for implications for econometric estimation of residual demand curves.

[57] It should be remembered that the Lerner Index for every fringe firm is equal to zero.

[58] Actually, the HHI in the formula is a truncated version of the true HHI because it excludes the fringe firms. The true HHI would include squared shares of all fringe firms.

[59] For more on the use of the HHI in merger analysis, see, e.g., Ordover and Willig [1983] and Ordover, Sykes, and Willig [1982]. The March 1983 issue of the *California Law Review* contains many useful articles discussing the 1982 Merger Guidelines and the role of the HHI. For a critique, see Farrell and Shapiro [1990].

discrete reduction in the incumbent's sales.[60] For example, if a market is perfectly contestable, entry occurs whenever the incumbent sets prices that yield a positive profit in the absence of entry. This can be taken to mean that the incumbent is facing, as it were, a perfectly elastic demand at that price vector. On the other hand, that same incumbent would be facing a much less elastic demand if the prospective entrant were to operate on the assumption that the incumbent would stick to the pre-entry vector of outputs. The US Department of Justice Merger Guidelines seem to adopt an expectational model that is close to the contestability framework. They inquire about hypothetical entry if current prices were elevated by a target amount and remained there for two years. Consequently, when there is a potential for entry, the Lerner Index at the profit-maximizing vector of the incumbent's actions (prices) depends on the specification of post-entry rivalry. This specification will, in turn, determine whether or not entry will occur in the first place.

### 3.1.4. Cross-product effects

The elasticity of the residual demand for a product depends on how prices of substitutes and complements adjust to changes in the price of the product in question. To see what is involved, consider the case of a multiproduct firm that is dominant in a number of markets. In setting the optimal price for product i, this firm will not only consider the usual elasticity of market demand for the product but also the effects of changes in the price of product i on the net revenues, $(p^j - MC^j)$, $j \neq i$, from its other products. If the increase in the price of product i stimulates the demand for other products that are priced above costs, the incumbent firm may elevate the price of product i by more than it would if it were not selling related products.[61]

Even if the incumbent does not control competing substitute products, it may conjecture—as it would in a price-leadership model—that as it raises its price other firms will follow suit, at least to some extent. In this case, predictions of the degree of market power based on

---

[60] Technically, the residual demand function is not differentiable at the entry-triggering price.
[61] For more on pricing by multiproduct firms, see, e.g., Baumol, Panzar, and Willig [1982].

the elasticity of the conventional market demand may underestimate the firm's ability to increase price profitably.

### 3.1.5. Intertemporal effects on the cost and demand side

Pindyck [1985] amply demonstrates that in dynamic models, the standard version of the Lerner Index could either over-or understate the actual degree of market power held by a firm. He defines a dynamic market as 'one in which price and production are intertemporally determined.' Examples of such markets include markets for exhaustible resources, markets in which firms learn by doing, markets with consumer-switching costs, and so on. Pindyck demonstrates that, in some cases, the instantaneous Lerner Index given in (3.3) can be 'rescued' as a measure of market power by appropriately measuring marginal cost. In his formula, the deviation of price is taken from the full marginal social cost ($FMC_t$) evaluated at the monopoly output level.

The important point to recognize in this adjustment is that the dominant firm also takes into account the intertemporal externalities. For example, when there are learning effects, the firm has an incentive to produce more in period t than would be forthcoming by equating $MR_t = MC_t$. The firm recognizes that an increase in current output has a beneficial effect on future marginal costs, so it will set $MR_t + \delta_t = MC_t$, where $\delta_t$ is the dollar value of future cost reductions (along the optimal path) from a small expansion in output at time t. In the case of learning by doing, Pindyck shows that the social value of the expansion is greater than the private value, meaning that the full marginal social cost is less than the private marginal cost of an expansion, i.e. $FMC_t < MC_t - \delta_t$. In this case, the usual Lerner Index would underestimate instantaneous market power in each period. In the case of exhaustible resources, on the other hand, the index would overestimate such power.

I have demonstrated that the Lerner Index, when calculated using an appropriate measure of marginal cost and the elasticity of residual demand, provides a measure of market power. How does the Lerner Index method compare to the approach that uses market shares to assess market power? This is the issue considered in the next subsection.

## 3.2. Market Share as an Index of Market Power

Among methods of assessing market power, there may appear to be a significant difference between the LP (Lerner Index) approach and the traditional reliance on market shares. The two can obviously be reconciled by simply scaling the 'market' share up or down to reflect the presence of immediate and potential competitors, the presence of the fringe, the existence of barriers to entry or expansion, and so on. And, of course, the scope of the market can be adjusted to include substitute products so as to make market share a meaningful predictor of market power, as I discuss below.[62] Thus, since the market share approach requires that the market first be defined, various necessary adjustments can come in at the market definition stage.

It is possible to adduce various separate reasons that market shares ought to merit separate scrutiny in the process of analyzing market power. The first is articulated in a recent paper by Shepherd [1988]. He writes that '[m]onopoly effects usually begin to be significant as market shares [held by a dominant firm] rise above 15-20 percent, as shown by research on the [Market Share-Rate of Return] function. By 40-50 percent, the effects are usually quite strong,' pp. 20-21). The firm holding such shares could possibly be 'capable of eliminating any small rival it chooses, at any given time, or possibly all the rivals. It may choose not to do so from moment to moment, but its relative power to do so exists...' (*Ibid.*, pp. 8-9). According to this view, large market shares in themselves signify market power. Furthermore, the ability to engage in anticompetitive conduct is positively related to market share.

The sources of the positive correlation between market share and (excess) rate of return are subject to debate in theoretical and empirical literatures. For example, the 'Chicago School' and Demsetz explain the perceived correlation by superior efficiency: i.e., higher market shares accrue to more efficient firms.[63] But even if market share bestows the ability to eliminate competitors or exclude entrants, it does not follow

---

[62] Ordover and Willig [1986] discuss the necessary analytics in the context of 'world' versus 'domestic' market definitions for merger analysis.

[63] For more on this see Gilbert [1989] and the survey by Schmalensee [1989]. Scherer [1980, ch. 9] contains a critical review of pre-1980 regression studies.

that it provides the incentive to do so. In many market scenarios, it may not be profitable for a dominant firm to eliminate some or all of its rivals.[64] Consequently, the likelihood of anticompetitive conduct by firms with large market shares is smaller than one might surmise from considering only their ability to harm competitors and competition.

The second reason for relying on market shares as an index of market power stems from the difficulties inherent in estimating the pertinent residual elasticities of demand, as compared to estimating the elasticity of demand for the whole market.[65] Quite, generally, the smaller the market share of the firm, the higher is the residual elasticity of demand facing it. Knowing the elasticity of market demand and firm's share, we can make some inferences regarding the elasticity of demand facing it.[66]

When residual elasticities, or even market elasticities, cannot be easily estimated from the data, market shares in a 'reasonably' constructed market could possibly serve as indices of market power. In fact, the essence of the LP program for estimating market power lies in a series of 'adjustments' that have to be made in the traditional determination of markets and market share so as to better capture the role played by the elasticity of residual demand in constraining the ability of the dominant firm to raise prices above the competitive level.

To see what is involved, assume the fact-finder determined that milk is the relevant antitrust product market[67] and that Latte Farms, Inc., has 65 percent of the milk market. The remaining 35 percent of the milk comes from small, price-taking farms. The prevailing price of milk is $p°$ per gallon. Assume that if Latte were to raise the price slightly above $p°$ the milk market would be overwhelmed by the additional output from the fringe farms. Under those circumstances, Latte cannot succeed in driving the price of milk above $p°$ unless it succeeds in relaxing the constraint that kicks in when the price reaches $p°$. Obviously,

---

[64] Economists have constructed models in which the dominant firm in fact encourages competition as assurance to its buyers that they will not be 'overexploited' in the future.

[65] See below for the problems of estimating profits and price-cost margins.

[66] In recent years, industrial organization economists and econometricians have devised very powerful methods that in some circumstances permit direct estimation of the residual elasticities facing a firm (or a group of firms). These techniques are reviewed by Bresnahan [1989]. See also my discussion in subsection 4.

[67] Meaning that the demand for milk is rather inelastic, there being few good substitutes for milk.

though, Latte can still have market power in the sense that $p°$ could be significantly above marginal cost (or average cost).

How can we tell whether the elasticity of the fringe's supply is high?[68] One possibility is that the fringe firms are located elsewhere and are exporting milk into Latte's territory (i.e., Latte's territory is a potential geographic market). Then, according to LP, a small increase in price above the prevailing level would induce the fringe to divert sales from other markets into Latte's territory. Consequently, the appropriate market in which Latte's share could be calculated encompasses not only the current volume of imports (i.e., the fringe's sales in Latte's market) but also the total output of the fringe firms. In other words, Latte's share should be calculated in the 'world market.' Obviously, that share would most likely be less than Latte's 65 percent share at home.

A moment's reflection suggests, however, that whether we calculate Latte's share in the narrow market or in the 'world' market should not affect our assessment of whether Latte has market power. Broadening the geographic market to include the 'world' output of the fringe firms has the effect of lowering the elasticity of supply used to deflate market share in the Lerner Index. Which market share is a better predictor of market power depends upon our intuition about the elasticity of the fringe supply into Latte's market. If that elasticity is very large, the broad market seems appropriate.

We have seen that the market share approach and the elasticity approach can be partly reconciled through the process of adjustments in the size and scope of the market.[69] It is not immediately obvious, however, how two approaches relate to antitrust analysis that concerns itself squarely with conduct and not market power *per se*. To illustrate, assume that at a fully competitive level, a dominant firm faces a substantial demand elasticity because of intense rivalry from another firm. If the dominant firm could succeed in inducing its rival's exit, the elasticity of demand facing it would be substantially reduced, enabling it to elevate price. Since firms might be willing to expend resources to reduce the pertinent elasticities, antitrust inquiry should focus not only on the elasticity of demand but also on the consequences of firm

---

[68] My discussion here follows Ordover and Willig [1981b]. See also Brennan [1982].

[69] Another round of adjustments may entail the use of capacities rather than actual sales, as proposed by Landes and Posner. For the pitfalls see Schmalensee [1982] and Ordover and Willig [1983].

conduct on the elasticity facing it—i.e., inquiry should be made into the marginal productivity of a dollar spend on elasticity-reducing activities. In the example constructed above, one line of inquiry would thus focus on Latte's ability and incentives to weaken the constraint at p° through its actions and to forestall an increase in that elasticity with a goal of maintaining its market share and the price at p°.

Formula (3.1) also provides some support for the use of current market shares in merger analysis to assess the likely competitive effects of a merger. From the formula we see that a merger between two firms reduces the residual elasticity of demand facing the combined entity, hence increases market power. However, the formula can both underestimate the reduction (because the industry may become more collusive post-merger, see Ordover, Sykes, and Willig [1982]) and overestimate the reduction (because the non-merging firms are likely to expand output in response to the merger, see Farrell and Shapiro [1990]). The point made here, then, is that conduct-oriented analysis takes the current levels of market share and market power as the starting point and inquires into the consequences of the complained-of conduct, examining the likely increment in market power and the reduction in the speed with which this power would otherwise be eroded (as captured in Schmalensee's formula for the dynamic measurement of the cost of monopoly).

### 3.3. Profit

Excess economic profit is yet another index of market power. By definition, a competitive firm earns only a normal rate of return on its assets in long-run equilibrium, while a firm with market power enjoys above-normal profit.[70] If economic profit could be easily measured, it could be used to ascertain a firm's market power. In fact, excess long-run profit (or rate of return on assets) is a superior index, as compared with the Lerner Index, in those market situations where prices must remain above marginal costs if a firm is to earn a normal rate of return.

It can be shown that excess profit (or an excess rate of return) is linked to the Lerner Index in the following way:

$$\rho = b/[(S \cdot L) - (S - 1)],$$

---

[70] Supracompetitive profits should be distinguished from efficiency rents, which accrue to firms with superior assets.

where $\rho$ is the excess rate of return, and L is the Lerner Index. The parameter b is $P/(k/q)$, where P is the price and $(k/q)$ is the capital-output ratio. S is the ratio of average cost to marginal cost at equilibrium output, a measure of economies of scale. For a variety of reasons the parameters b and S are difficult to measure with precision. The expression for $\rho$, therefore, may not be of much value in assessing market power. Nonetheless, when there are pervasive economies of scale the Lerner Index overestimates market power. See Schmalesee [1979, 1982] and Ordover, Sykes, and Willig [1982].

Another pitfall involves the attempt to infer the value of $\rho$—and thus economic profit—from estimates of the *accounting rate of return*.[71] The misuse of the accounting rate of return as a proxy for economic profit has been widely debated. Fisher and McGowan [1983] demonstrated that the choice of depreciation policies can cause a divergence between accounting and economic profit. See also Fisher, McGowan, and Greenwood [1983, ch. 7], and Schmalensee [1989] for added technical discussion. Some authors have, however, questioned the extreme position taken by Fisher, McGowan and Greenwood. Thus, Ravenscraft and Scherer [1987, pp. 17-18] conclude that accounting profit data can be used as proxies for economic profit as long as there are no systematic errors, or errors correlated with 'the phenomenon . . . under investigation.' Kay [1987] argues that the economic rate of return can be related to the constructed variable, which he calls the accounting rate of return. That variable is distinct from the accounting rate of profit and can be unbiasedly used to assess the extent of market power held by a firm, in some instances.

I have reviewed various methods for assessing market power. As the term indicates, that power has to be assessed in a relevant antitrust market. In the next subsection, I briefly discuss the methods for constructing antitrust markets.

## 4. MARKET DEFINITION

An *antitrust market* (product and geographic) is best defined as a set of constraints on the ability of a firm (or group of firms) to exercise

---

[71] The accounting rate of return is accounting profit divided by stockholders' equity or total capitalization.

market power. Such a constraint could be a substitute product or another firm. When these contraints are powerful, a firm may not be able to raise the price of its product above the competitive level. For example, if a single seller of apples were to attempt to increase its price, consumers would switch to other sellers' apples. These other sellers constitute a competitive constraint. This does not mean, however, that if all sellers of apples could form a perfect cartel, they would not be able profitably to elevate the price significantly above the competitive level for a transitory period of time.

For such a price increase to be profitable, the usual market elasticity of demand for apples would have to be low enough to make apples a plausible antitrust market.[72] Conversely, if apple buyers would switch in sufficient numbers to other fruits in response to a price increase, 'apples' would not constitute the relevant product market, but 'fruits' would. Basic economics tells us that as we expand the scope of the market, the elasticity facing the aggregate of the firms within that market will fall.[73] At some point, the elasticity will become small enough to satisfy the market-definitional requirements.[74]

One aspect of the procedure just outlined is apt to raise serious methodological problems. This is the choice of the benchmark for the initial estimation of elasticity. It arises from the fact that an imperfectly competitive firm faces elastic demand at its profit maximum, so a price increase beyond this point must lower profit. Indeed, at a profit-maximizing point, any firm faces an elastic residual demand curve. Since the estimated elasticity is high, a fact-finder may conclude that the relevant product market includes other products as well, which would dilute the firm's market share.[75] Plainly, if the pertinent elasticity were evaluated at the 'competitive' level, $p = MC$, it could be significantly lower than it would be at the imperfectly competitive

---

[72] This definition essentially adopts the methodology of the US Department of Justice Merger Guidelines issued in 1982 (rev. 1984).

[73] Note that as the market is expanded to include non-identical substitutes (say, apples and oranges), the definition of market elasticity is no longer unambiguous. See Landes and Posner [1981], Schmalensee [1982], Kaplow [1982], and Ordover and Willig [1981].

[74] One unsatisfactory outcome of this process is that once that market has been defined, the firms and products not in the market are assigned zero market shares in the relevant calculations, despite the fact that they do exert some constraint on the firms within the market.

[75] United States v. E. I. du Pont de Nemours & Co., 351 U.S. 377 [1956].

optimum. At that lower price, a narrower market definition might be mandated. Unfortunately, there may be no price and quantity data that would permit confident out-of-sample estimation of the pertinent elasticities. The proposed focus on the effects of conduct on the elasticity of demand facing a firm (or a group of firms) overcomes some of the problems just described.

Various techniques have been developed for the purposes of constructing antitrust markets. Practitioners disagree on which methods are more effective. Two approaches have gained some prominence in recent years.[76] The first approach has been outlined by Stigler and Sherwin [1985]. It relies on the examination of correlations between price movements of different products, or between the same products sold in different geographic areas. A high correlation coefficient indicates that the two products (or geographic areas) are linked by the forces of demand-side and supply-side arbitrage, and hence fall within the same market.[77] The shortcoming of the correlation approach is that it does not focus directly on any constraints on the exercise of market power other than arbitrage. Arbitrage may temper the exercise of market power, but not necessarily eliminate it.

An alternative approach is to estimate directly those demand elasticities that capture whatever constraints there may be on the exercise of market power. This approach has been advanced by Scheffman and Spiller [1987], Baker and Bresnahan [1988], Baker [1987] and is discussed at length by Bresnahan [1989]. The goal of this approach is to identify a region and set of products within which demand is highly elastic, but outside of which demand is inelastic. The approach is directly responsive to the US Department of Justice Merger Guidelines' definition of a market. Under the Guidelines, an antitrust market is defined as a group of firms or products that would, if they formed a perfect cartel, profitably raise prices above the current level for a non-transitory period of time. Hence, the Guidelines require one to estimate

---

[76] See Ordover and Wall [1989] for a more detailed but non-technical description. A technical comparison can be found in Slade [1987] and Baker [1987].

[77] Slade's version of the approach is to regress the price of a product at time t against its lagged past prices and other variables (such as labor costs, for example) and then test for a gain in the predictive power of the regression by introducing present and past prices of possible substitutes. A significant gain in predictive power implies that the two products are in the same market. See Slade [1987].

the elasticity of *market* demand facing the putative cartel around some benchmark price level.

The approach is not free of shortcomings, however. First, it requires detailed information on demand and supply shifters—i.e., shocks to demand and supply that affect only some products but not others. Such information is not readily available in many situations. Unfortunately, without such information it may be impossible to estimate the key parameters. Second, the estimated residual elasticities can differ from the notional elasticities used by firms in their decision-making. As a result, the estimated elasticities may not correctly measure the firms' incentives to raise price above the competitive level. Third, merger analysis using this approach is made difficult (albeit not impossible) by the fact that the estimation process proceeds on the assumption of a constant degree of market rivalry among the existing firms, irrespective of their number.

This brief summary indicates that the market for market definition methods is highly competitive. However, great strides have been made in our understanding what types of techniques (what types of data) will have to be developed in order to reduce some of the fuzziness surrounding the issue of market definition.

## 5. CONCLUDING COMMENTS

This section laid out the basics of economic foundations of competition policy. I made it clear that when the competitive process fails, short-term and long-term resource allocation can suffer. I also indicated that the effectiveness of the competitive process cannot be readily assessed by such measures as the number of firms in the market, the distribution of market shares, price/cost margins, or even profit rates, because these also reflect returns to superior efficiency, business acumen, and innovative success. The discussion suggested that the analytic assessment should first focus on the assets that in the particular setting can plausibly generate market power. The next step in the assessment should address the incentives, rather than the diffused abilities, of a firm (or group of firms) to undermine the competitive process.

The competitive process benefits consumers insofar as it acts as a constraining force. But prices to consumers are not the only variable of concern in the *economic* calculus. Production costs, product quality, and allocation of resources towards R&D are also pertinent. These additional policy concerns are often best served by business strategies that at first may seem inimical to the goals of free and unhampered competition. That such strategies are not always welfare-reducing in the long run has been demonstrated by the growing literature on vertical restraints, joint ventures, and technology licensing.

What is still missing from the current foundations of antitrust economics is the kind of analytic certitudes that underlie the traditional structure-conduct-performance paradigm and now inform the Chicago view of markets. Unfortunately, the strategic approach to industrial organization, which underlies antitrust economics, does not readily generate simple generalizations.[78]

## BIBLIOGRAPHY

[1] Baker, J., (1987), 'Why Price Correlations do not Define Antitrust Markets: On Econometric Algorithms for Market Definition,' Working Paper No. 149, US Federal Trade Commission, Washington, DC.
[2] Baker, J. and Bresnahan T. F. (1988), 'Estimating the Residual Demand Curve Facing a Single Firm,' *International Journal of Industrial Organization*, **6**, 283–300.
[3] Baldwin, W. L. and J. T. Scott, (1987), *Market Structure and Technological Change*, Fundamentals of Pure and Applied Economics Series, J. Lesourne and H. Sonnenschein (eds), Chur, Switzerland: Harwood Academic Publishers.
[4] Baumol, W. Panzar, J. and Willig R. D. (1982), *Contestable Markets and the Theory of Industry Structure*, New York: Harcourt, Brace, Jovanovich.
[5] Baumol, W. Panzar J. and Willig R. D. (1986), 'On the Theory of Perfectly Contestable Markets,' in *New Developments in the Analysis of Market Structure*, J. Stiglitz and F. Mathewson (eds), Cambridge: MIT Press.
[6] Baumol, W. and Willig R. D. (1986), 'Contestability: Developments Since the Book,' *Oxford Economic Papers*, **38**, Supplement, 9–36.
[7] Bork, R. (1966), 'The Legislative Intent and the Policy of the Sherman Act,' *Journal of Law and Economics*, **9**, 7–48.
[8] Bork, R. (1978), *The Antitrust Paradox*, New York: Basic Books.
[9] Brennan, T. J. (1982), 'Mistaken Elasticities and Misleading Rules,' *Harvard Law Review*, **95**, 1849–56.
[10] Bresnahan, T. F. (1989), 'Empirical Studies of Industries with Market Power,' *Handbook of Industrial Organization*, Willig R. D. and Schmalensee R. (eds), Amsterdam: North Holland vol. II, ch. 17.

---

[78] See, e.g., Tirole's [1988] textbook and a spirited exchange between Fisher [1989] and Shapiro [1989] in the *Rand Journal of Economics*.

[11] Comanor, W. S. and Smiley, R. H. (1975), 'Monopoly and the Distribution of Wealth,' *Quarterly Journal of Economics*.
[12] Demsetz. H. (1973), 'Industry Structure, Market Rivalry, and Public Policy,' *Journal of Law and Economics*, **16**, 1-10.
[13] Farrell, J. and Shapiro, C. (1990), 'Horizontal Mergers: An Equilibrium Analysis,' *American Economic Review*, **80**, 127-142.
[14] Fisher, F. (1985), 'The Social Costs of Monopoly and Regulation: Posner Reconsidered,' *Journal of Political Economy*, **93**, 410-16.
[15] Fisher, F. and McGowan, J. (1983), 'On the Misuse of Accounting Rates of Return to Infer Monopoly Profits,' *American Economic Review*, **73**, 82-97.
[16] Fisher, F. McGowan J. and Greenwood, J. (1983), *Folded, Spindled, and Mutilated: Economic Analysis of US V. IBM.*, Cambridgee: MIT Press.
[17] Fox, E. M. (1982), 'The New Merger Guidelines-A Blueprint for Microeconomic Analysis,' *Antitrust Bulletin*, **27**, 519.
[18] Fox, E. M. (1986), 'The Politics of Law and Economics in Judicial Decision Making: Antitrust as a Window,' *New York University Law Review*, **61**, 554-588.
[19] Fox, E. M. (Winter 1988), 'Antitrust, Economics and Bias,' *Antitrust*, **2**, 6-10.
[20] Fox, E. M. and Sullivan, L. A. (1987), 'Antitrust-Retrospective and Prospective: Where Are We Coming From? Where Are We Going?' *New York University Law Review*, **62**, 936-988.
[21] Fudenberg, D. and Tirole, J. (1987), 'Understanding Rent Dissipation: On the Use of Game Theory in Industrial Organization,' *American Economic Review: Papers and Proceedings*, **77**, 176-183.
[22] Gaskins, D. (1971), 'Dynamic Limit Pricing: Optimal Pricing Under Threat of Entry,' *Journal of Economic Theory*, **2**, 306-322.
[23] Geroski, P. Gilbert R. J. and Jacquemin A. (1990), *Barriers to Entry and Strategic Competition*, in Fundamentals of Pure and Applied Economics, J. Lesourne and H. Sonnenschein (eds), New York: Harwood Academic Publishers.
[24] Gilbert, R. J. (1989), 'The Role of Potential Competition in Industrial Organization,' *Journal of Economic Perspectives*, **3**, 107-127.
[25] Gilbertm R. J. (1989), 'Mobility Barriers and the Value of Incumbency,' *Handbook of Industrial Organization*, Schmalensee R. and Willig R. D. (eds), Amsterdam: North Holland, vol. I, ch. 8.
[26] Hawk, B. (1986), *United States, Common Market and International Antitrust: A Comparative Guide*, 2nd ed., New York: Law and Business, Inc.
[27] Kaplow, L. (1984), 'The Patent-Antitrust Intersection: A Reappraisal,' *Harvard Law Review*, **97**, 1813-1892.
[28] Kaplow, L. (1982), 'The Accuracy of Traditional Market Power Analysis and a Direct Adjustment Alternative,' *Harvard Law Review*, **8**, 1817-48.
[29] Kay, J. A. (1987), 'Assessing Market Dominance Using Accounting Rates of Profit,' in *The Economics of Market Dominance*, Hay D. and Vickers J. (eds), Oxford: Basil Blackwell, 129-142.
[30] Koenker, R. W. and Perry M. K. (1981), 'Product Differentiation, Monopolistic Competition, and Public Policy, *Bell Journal of Economics*, **12**, 217-231.
[31] Lande, R. (1988), 'The Rise and (Coming) Fall of Efficiency as the Ruler of Antitrust,' *The Antitrust Bulletin*, **33**, 429-465.
[32] Landes, W. M. and Posner, R. A. (1981), 'Market Power in Antitrust Cases,' *Harvard Law Review*, **94**, 937-96.
[33] Meuller, D. (1986), *Profits in the Long Run*, Cambridge: Cambridge University Press.
[34] Nozick, R. (1974), *Anarchy, State and Utopia*, New York: Basic Books.

[35] Ordover, J. A. and Baumol, W. (1988), 'Antitrust and High-Technology Industries, *Oxford Review of Economic Policy*, **4**, 13-34.
[36] Ordover, J. A. and Wall D. (1989), 'Understanding Econometric Models of Market Definition', *Antitrust*, **3**, 20-25.
[37] Ordover, J. A. and Willig R. D. (1981a), 'An Economic Definition of Predation: Pricing and Product Innovation', *Yale Law Journal*, **91**, 8-53.
[38] Ordover, J. A. and Willig R. D. (1981b), 'Market Power and Market Definition,' Memorandum for the ABA Section 7 Clayton Act Committee Project on Revising the Merger Guidelines.
[39] Ordover, J. A. and Willig, R. D. (1983), 'The 1982 Department of Justice Merger Guidelines: An Economic Assessment,' *California Law Review*, **71**, 535-574.
[40] Ordover, J. A. and Willig R. D. (1986), 'Perspectives on Mergers and World Competition,' in *Antitrust and Regulation*, R. Greison (ed.), Lexington: Lexington, 201-218.
[41] Ordover, J. A. Sykes, A. O. and Willig, R. D. (1982), 'Herfindahl Concentration, Rivalry, and Mergers, *Harvard Law Review*, **95**, 1857-74.
[42] Pindyck, R. S. (1985), 'The Measurement of Monopoly Power in Dynamic Markets,' *Journal of Law and Economics*, **28**, 193-222.
[43] Philips, L. (1983), *The Economics of Price Discrimination*, Cambridge: Cambridge University Press.
[44] Posner, R. A. (1975), 'The Social Cost of Monopoly and Regulation,' *Journal of Political Economy*, **83**, 807-27.
[45] Pynchon, T. (1973), *Gravity's Rainbow*, Viking Press, New York.
[46] Ravenscraft, D. J. and Scherer, F. M. (1987), *Mergers, Sell-Offs, and Economic Efficiency*. Washington D.C.: Brookings Institution.
[47] Reinganum, J. (1989), 'The Timing of Innovation: Research, Development and Diffusion,' *Handbook of Industrial Organization*, Willig, R. D. and Schmalensee R. (eds), (Amsterdam: North Holland) vol. 1, ch. 14.
[48] Robinson, J. (1933), *The Economics of Imperfect Competition*, London: Macmillan.
[49] Salinger, M. A. (1984), 'Tobin's q, Unionization, and the Concentration-Profits Relationship,' *Rand Journal of Economics*, **15**, 159-170.
[50] Scheffman, D. T. and Spiller, P. (1987), 'Geographic-Market Definition Under the U.S. Department of Justice Merger Guidelines,' *Journal of Law and Economics*, **30**, 123-148.
[51] Scherer, F. (1980), *Industrial Market Structure and Economic Performance*, 2nd ed., Boston: Houghton Mifflin.
[52] Schmalensee, R. (1979), 'On the Use of Economic in Antitrust: the ReaLemon Case,' *University of Pennslyvania Law Review*, **127**, 994-1050.
[53] Schmalensee R. (1982), 'Another Look at Market Power,' *Harvard Law Review*, **95**, 1789-816.
[54] Schmalensee, R. (1989), 'Inter-Industry Studies of Structure and Performance,' *Handbook of Industrial Organization*, Schmalensee, R. and Willig, R. D. (eds), Amsterdam: North Holland.
[55] Schwartz, M. (1986), 'The Nature and Scope of Contestability Theory,' *Oxford Economic Papers*, Supplement, **38**, 37-57.
[56] Shapiro, C. (1986), 'Exchange of Cost Information in Oligopoly,' *Review of Economic Studies*, **53**, 433-46.
[57] Shepherd, W. G. (1984), 'Contestability vs. Competition,' *American Economic Review*, **74**, 572-87.
[58] Shepherd, W. G. December 29, 1988, 'The Process of Effective Competition,'

presented at the American Economic Association meetings, December 29, 1988.
[59] Slade, M. July 1987 'Another Look at Market Definition in Antitrust,' mimeo, University of British Columbia.
[60] Stigler, J. (1968), *The Organization of Industry*, Homewood Ill.: Irwin.
[61] Stigler, J. and Sherwin, R. (1985), 'The Extent of the Market,' *Journal of Law and Economics*, **28**, 555-585.
[62] Stiglitz, J. (1988), 'Technological Change, Sunk Costs, and Competition,' *Brookings Papers on Economic Activity*, Special Issue on Microeconomics, 3/1987, 883-937.
[63] Thomas, L.G. (March 1989), 'Asymmetries in Entry Competence: A Revisionist Analysis of the RTE Cereals Industry,' Working Paper 89-14, Columbia University Graduate School of Business.
[64] Tirole, J. (1988), *The Theory of Industrial Organization*, Cambridge: MIT Press.
[65] qarian, H. R. (1989), 'Price Discrimination,' *Handbook of Industrial Organization*, Schmalensee R. and Willig, R. D. (eds), Amsterdam: North Holland, vol. 1, ch. 10.
[66] Vickers, J. and Hay D. (1987), 'The Economics of Market Dominance,' in *The Economics of Market Dominance*, Vickers J. and Hay D. (eds), Oxford: Basil Blackwell, pp. 1-60.
[67] Whinston, M. D. (June 1989), 'Tying, Foreclosure, and Exclusion,' NBER Working Paper No. 2945.
[68] Williamson, O. E. 'Economics as an Antitrust Defense: The Welfare Trade-off', *American Economic Review*, **58**, 18-36.
[69] Williamson, O. E. (1977b), 'Economics as an Antitrust Defense Revisited,' *University of Pennsylvannia Law Review*, **125**, 699.

# United States Antitrust Policy: Issues and Institutions

WILLIAM S. COMANOR*
*University of California, Santa Barbara*

## 1. INTRODUCTION

The United States policy of promoting competition is one hundred years old. In 1890, the Sherman Antitrust Act was passed, and its two major provisions still remain the basis of current law. The essential reason for this longevity is the breadth and sweep of the statutory language: Section 1 declares unlawful contracts and combinations 'in restraint of trade,' while Section 2 makes it illegal to 'monopolize, or attempt to monopolize.' But the statute defines neither term and federal judges therefore have the task of defining the nation's competition policy.

In the years and decades that followed the passage of the Sherman Act, new laws appeared to support or limit the original provisions, but these changes lay within the context of the original law. The most important appeared in 1914 when Congress passed the Clayton Act. Section 7 of that law prohibits acquisitions that 'substantially ... lessen competition, or ... tend to create a monopoly.' In 1950, the Clayton Act was extended to apply to asset as well as stock acquisition.

Also in 1917, Congress enacted the Federal Trade Commission Act which created the Commission and provided it with the power to prohibit 'unfair methods of competition.' While that phrase might seem to provide wide latitude to regulate the economy, in effect it provided similar authority to that of the Sherman and Clayton Acts.

While there are other operative provisions, these four set the basic framework within which most policy actions are taken. They are

---
*I am very grateful for the helpful comments and suggestions of John S. Willey.

enforced by two separate agencies, which have similar functions. First, a division in the US Department of Justice plays a prosecutorial role and brings prospective antitrust violators before a US District Court, with appeals taken directly to the Supreme Court. Final outcomes thereby turn on enforcement decisions in determining which cases are brought, as well as on judicial decisions.

The enforcement policies of the Federal Trade Commission are more cumbersome. With majority approval of this five-member body, the Commission staff brings civil but not criminal complaints before an Administrative Law Judge, who is also employed by the Commission. A full trial is heard with appeals taken to the Commission. These decisions can then be appealed to US Courts of Appeal and the Supreme Court. As a result, there are more steps required before reaching final outcomes in Commission actions.

Whatever the costs of having dual prosecutorial agencies, with its need for a detailed clearance procedure to allocate cases between them, there are important benefits as well. Among the most important are the greater breadth of opinion on appropriate policy standards. While generally there is agreement on policy directions, this is not always the case. Even if one side sees no competitive harms in some practice or circumstances, the other agency may have a different view so that the issue is more likely to be adjudicated. With two decision-makers rather than one, there is more prospect for an active policy.

While both government agencies have played a major role in the development of the policy, it is also called upon by private parties. Private litigants can sue under both the Sherman and Clayton Acts, and are rewarded with treble damages if they prevail. So important is this vehicle for antitrust enforcement that about 90 percent of all antitrust cases are brought by private litigants where neither federal agency is a direct party.[1] Whatever the litigants' objectives, such cases often have major effects. They can also be appealed to higher courts, and some Supreme Court decisions of private cases have altered the course of policy development. Particularly in recent years, when both federal agencies have adopted a minimalist approach in some areas, private antitrust litigation has continued to provide sustenance to antitrust enforcement.

While cases brought by the federal enforcement agencies cover all

---

[1] Lawrence J. White, (ed.), (1988) *Private Antitrust Litigation, New Evidence, New Learning*, Cambridge: MIT Press, p. 4.

policy areas, those brought by private litigants tend to be concentrated in relatively few. Their cases are typically concerned with horizontal price fixing, refusals to deal, and tying or exclusive dealing. Together, these three allegations appear in over half of all private complaints.[2] Nearly two-thirds of all private plaintiffs are either competitors or distributors of the defendants.[3] Such firms may be disadvantaged by antitrust violations, and private actions are a likely result.

## 2. A BRIEF REVIEW

After a slow start, which carried through most of the 1890s, an active policy was pursued during the administration of President Theodore Roosevelt which took office in 1901. An Antitrust Enforcement Division was created in the Department of Justice and a number of new cases were filed. The most important of these led to the two landmark decisions of 1911 that dissolved single firm monopolies in the oil and tobacco industries.[4] From that point on, the enormous potential of antitrust for affecting the structure of the American economy was established.

By the 1960s and into the 1970s, a body of antitrust law and doctrine had developed, which commanded the support of both major political parties.[5] As a result, in this era, one saw few changes in policy directions linked directly to partisan politics. Instead, one found both strong and weak, effective and ineffective, active and inactive antitrust regimes under both political parties.[6]

The essential purpose of antitrust policy during that period was the prevention or limitation of monopoly (or market) power. As expressed in a famous Supreme Court decision of the time, 'monopoly power is the power to control prices or exclude rivals'[7] and enforcement officials saw it their task to impede the expansion of market power where

---

[2] *Ibid.*, p. 6.

[3] *Ibid.*, p. 8.

[4] *United States v. Standard Oil Company of New Jersey*, 221 US 1 (1911); and *United States v. American Tobacco Company*, 221 US 106 (1911).

[5] For a valuable history of US antitrust into the 1960s, see Richard Hofstadter, 'What Happened to the Antitrust Movement?' in Richard Hofstadter, *The Paranoid Style in American Politics and Other Essays*, New York, (1965) pp. 188-237.

[6] See William S. Comanor, (Winter 1982) 'Antitrust in a Political Environment,' *Antitrust Bulletin*, Vol. XXVII, pp. 733-752.

[7] *United States v. Dupont and Co.*, 351 US 377, 391.

possible and prevent the exploitation of any market power which remained. Throughout, there was little attention paid directly to questions of efficiency.

The leading decisions of the era were *Schwinn* and *Von's Grocery*,[8] which set the high-water marks for antitrust enforcement in their respective areas. The *Schwinn* decision concerned non-price vertical restraints and ruled that, so long as all indicia of product ownership had passed to the distributor, these restraints were illegal *per se.* Similarily, the *Von's* decision prevented a merger between two supermarket chains which together had less than 8 percent of the relevant local market. From this decision, many believed that all horizontal mergers were prohibited.

Also in this era, the Department of Justice brought its massive monopolization case against IBM, which dragged on for over a decade. The matter was finally discharged only after a lengthy discovery or fact-finding process which did not substantiate many of the government's charges, and also after market circumstances had substantially changed.[9] Other important monopolization cases of that period were the Department of Justice suit against AT&T and the FTC action against the four leading manufacturers of breakfast cereals. Like the IBM case, the Commission eventually abandoned the Cereals case although after an Administrative Law Judge had rendered a decision.[10] On the other hand, the suit against AT&T had major consequences and led to a substantial restructuring of the telecommunications industry.[11] These cases are noteworthy not only because of their very different results but also because they were so rare. Even in an era of active enforcement, these were few large cases charging monopolization and their outcome was always uncertain.

Throughout that period, a latent controversy lay just below the surface of policy choices, which was the possible conflict between the

---

[8] *United States v. Arnold, Schwinn and Co.*, 388 US 365 (1967) and *United States v. Von's Grocery Co.*, 384 US 270 (1966).

[9] See F. M. Fisher *et al.* (1983), *Folded, Spindeled, and Mutilated: Economic Analysis and US v IBM*, MIT Press.

[10] See the second reference in note 36 below.

[11] See Roger Noll and Bruce M. Owen, 'The Anticompetitive Uses of Regulation: United States v. AT&T, 'John E. Kwoka and Lawrence J. White, (1989) *The Antitrust Revolution*, pp. 290–337; and Peter Tenin *et al.* (1987), *The Fall of the Bell System*, Cambridge University Press.

prevention of market power and the achievement of economic efficiency. Although the former was the stated task of antitrust, critics argued that was misguided and one should deal directly with consumer welfare and economic efficiency. Too often, they stated, antitrust diminished rather than enhanced economic efficiency. The response was that a market power standard led generally to improved consumer welfare and greater economic efficiency, although not necessarily in every instance. Since it was too difficult to evaluate efficiency in individual cases, a market power standard was more workable, and led on average to desired results.[12]

The tide began to turn with the *Sylvania* decision of 1977 that reversed antitrust standards towards vertical restraints.[13] In that case, the earlier standards of the *Schwinn* decision were reversed, and the legality of the restraints now turned expressly on their effects on economic efficiency. This action opened the door to a new era of restrained antitrust enforcement.

The *Sylvania* decision established the importance of economic efficiency in antitrust decisions. While efficiency had been a component of the earlier focus on market power, through the general presumption that diminished market power led to enhanced efficiency, the *Sylvania* opinion brought efficiency concerns into antitrust decision-making on a case-by-case basis.

The issue of efficiency permeated the antitrust enforcement agencies from the start of the Reagan Administration in 1981. After some hesitancy, enforcement officials suggested outright that the appropriate test for an antitrust case was whether the action increased the sum of consumer and producer surplus.[14] In effect, they proposed a 'consumer welfare' standard in which a cost-benefit analysis would be applied to every case. Competitive effects were relevant only through their effects on the conventional criteria for economic efficiency. There was a major shift in enforcement standards which had an impact as well on judicial decisions. A revolutionary change had occurred.

---

[12] The clearest expression of an unabashed market power standard appears in Carl Kaysen and Donald F. Turner, (1959) *Antitrust Policy*, Harvard University Press.

[13] *Continental T. V. v. GTE Sylvania*, 433 US 36 (1977).

[14] Charles F. Rule, Asistant Attorney General, Antitrust Division, US Department of Justice, 'Antitrust, Consumers and Small Business,' Remarks at the 21st New England Antitrust Conference, November 13, 1987, pp. 3–4.

As might be expected, these changes brought forth a flurry of complaints.[15] Moreover, with the decline in federal enforcement, there has been a substantial increase in enforcement actions by state officials that rest on state antitrust provisions. However, with the concurrance of some judicial opinions, a new synthesis may be developing that stands between the simple market power standard of the 1960s and the more recently proposed efficiency standard. In what might be called an extended market power standard, attention is directed at possible anti-competitive effects, although the focus is explicitly on competition rather than merely on competitors. And also it is increasingly accepted that defendents have an efficiency defense available to them in which any such gains must be balanced against any anticompetitive effects that result from particular conduct.

Whatever the arguments for and against the revised policy standards, one result is apparent. Antitrust enforcement policies have become a partisan issue as they were not in the past, and we can expect to find enforcement policies depending more directly on which political party is in power. The earlier consensus on antitrust has been broken, and future policy directions will be affected by the political outcomes of the electoral process.

In the sections below, we consider current antitrust standards as they apply in four major policy areas and place them in the context of the extended debate over just what these standards should be. To be sure, some policy areas are neglected; there is no claim of completeness in so brief a review. Still, we consider those areas where policy effects have been greatest.

## 3. THE PROHIBITION OF COLLUSIVE BEHAVIOR

The antitrust prohibition of agreements among competitors has long been the most basic component of US antitrust policy. From the outset, there was wide agreement that collusion led directly to mono-

---

[15] See, for example, Robert H. Lande, (September 1982) 'Wealth Transfers as the Original and Primary Concern of Antitrust: The Efficiency Interpretation Challenged,' *Hastings Law Journal*, Vol. 34, pp. 67–151, and Eleanor Fox *et al.* (1989), *The Fall and Rise of Antitrust: the Law in its Second Century*, Greenwood Press: Westport, CT.

polistic pricing and practices with no offsetting gains. It was a clear offense.

So strong has been the goal to eliminate such conduct that it led to the concept of a *per se* offense. In this context, the Supreme Court stated that: 'there are certain agreements or practices which because of their pernicious effect on competition and lack of any redeeming virtue are conclusively presumed to be unreasonable and therefore illegal without elaborate inquiry as to the precise harm they have caused or the business excuse for their use.[16]

For over half a century, the concept of *per se* has been applied to flagrant forms of collusive behavior. Explicit price-fixing agreements, group boycotts, and certain market division arrangements are adjudicated under this test. What is required is simply to demonstrate the presence of these arrangements without the need to describe their competitive effects in the case at hand. With *per se* violations, no facts can be used to justify the proscribed conduct.[17]

Recently, however, there was some erosion of this concept as the Court permitted an efficiency defence even in this section of the law.[18] The context was the licensing of copyright musical compositions to radio and television stations. By acting collectively, copyright owners sought to monitor and prevent unauthorized use of their compositions. But competition among them was eliminated in the process. The two major organizations representing copyright owners negotiated blanket licenses with the leading television networks and distributed royalties to copyright owners according to the nature and use of their music as well as other criteria.

The courts had considerable difficulty with this situation. The Court of Appeals overturned the lower District Court decision, and the Supreme Court in turn overturned the appellate decision. The problem faced by all was that the arrangement clearly restrained competition among composers but also had evident economic benefits.

With the *Sylvania* decision only two years earlier, the Court settled

---

[16] *United States v. Northern Pacific Railway*, 356 US 1, 5 (1958). While this case offers the standard rationale for *per se*, the rule was suggested earlier in *United States v. Trenton Potteries Co. et al.*, 273 US 392 (1927).

[17] For an interesting discussion of the *per se* concept, see Thomas G. Krattennaker (1988) 'Per Se Violations in Antitrust Law: Confusing Offenses with Defenses', *Georgetown Law Journal*, Vol. 77, pp. 165–180.

[18] *Broadcast Music, Inc., v. Columbia Broadcasting System*, 44 US 1 (1979).

directly on an efficiency explanation. It acknowledged that 'individual sales transactions in this industry are quite extensive,' so that 'a middleman with a blanket license was a ... necessity if the thousands of individual negotiations, a virtual impossibility, were to be avoided.' The Court continued that' this substantial lowering of costs ... differentiates the blanket license from individual use licenses.' Although a price-setting arrangement, the Court ruled that it 'should be subjected to a more discriminatory examination under the rule of reason.'[19]

More recently, however, the Supreme Court reaffirmed its original position on price-fixing arrangements. In the *Maricopa* decision of 1982, the Court found that a maximum price-fixing scheme was illegal *per se*, even though it was part of a medical insurance plan that 'had saved patients and insurers millions of dollars.'[20] It reasserted the standard of *per se* illegality for price fixing agreements.

Other types of arrangements are examined under the 'rule of reason'. Using that criterion, the plaintiff must demonstrate not only that particular behavior has anticompetitive effects but also that such effects dominate any purported justifications. This requirement is typically met by defining a relevant market and describing the implications of specific behavior for competitive outcomes in that market. The legal tasks are more onerous; more evidence is required; and there is more doubt that the particular conduct will be prevented. At the same time, the extended investigation offers more opportunity to evaluate particular conduct and tailor policy decisions more closely to the situation at hand.

A critical issue therefore is what constitutes a price-fixing or collusive arrangement, and what evidence is required to demonstrate its presence. In particular, a major question is whether parallel conduct is sufficient to indicate tacit collusion.

Currently, the law is settled that parallel conduct, without more, does not violate the Sherman Act. What more is needed, however, is uncertain, although a typical requirement is some indication that uniform behavior would not persist without a definite agreement. For the most part, the focus has been on whether the available evidence is

---

[19] *Ibid.*
[20] *Arizona v Moricopa Country Medical Society*, 457 US 332, 342 (1982).

sufficient to demonstrate an agreement, whether tacit or explicit. And in this context, there is no unequivocal answer as to whether tacit collusion by itself and without more is proscribed.

Whatever the form of an agreement among rivals, and whether or not it extends beyond price-setting behavior, this matter concerns joint or coordinated decision-making.[21] This issue is relevant because jointly determined prices and quantities are different and indicate impaired efficiency as compared with those set independently. Therefore, while society's ultimate concern may be with the particular prices and quantities that are set, the policy deals instead with the process by which they are set.

An inherent problem is what policy standards should apply to collusive agreements between a small number of sellers out of many in a particular market. While a process standard requires condemnation, this type of agreement cannot affect market results so long as many firms are not party to the arrangement. This hypothetical situation poses the prospect of significant differences between standards that focus on market processes from those which deal with particular results.

This emphasis on process rather than results has lead Richard Posner to suggest that it 'is inconsistent with an effective antitrust policy. Many attempts to fix price may have negligible consequences, while much serious price-fixing may escape detection altogether because overt communication ... may not always be necessary to effectuate price fixing.'[22] He argues instead that it would be appropriate for 'a jury to find an agreement to fix prices if it is satisfied that there was a tacit meeting of the minds of the defendants on maintaining a non-competitive pricing policy.'[23] The trappings of an explicit agreement are not essential, as compared with a finding that economic factors suggest non-competitive results.

While this proposal has much appeal, the problem is one of enforceability. While the current policy prohibits direct dealings with

---

[21] For an interesting discussion of these issues, see John S. Wiley, (August 1988) 'Reciprocal Altruism as a Felony: Antitrust and the Prisoners' Dilemma,' *Michigan Law Review*, Vol. 86, pp. 1906–1928.

[22] Richard A. Posner, (1976) *Antitrust Law: An Economic Perspective*, Chicago: University of Chicago Press, p. 41.

[23] *Ibid.*, p. 72.

competitors, the corresponding prohibition under a broader concept of collusion would be that one should not behave interdependently. But so broad a prohibition is as difficult to enforce as it is to define.

Where the number of firms in a market is few, firms invariably account for the actions of their rivals in making their own decisions, and it strains credibility to suggest that they will not take such reactions into account because of antitrust prohibitions. In defining prohibitions, firms must understand what conduct is permitted as well as what is prohibited, and this proposal makes it far more difficult to describe permissible conduct. In response, Posner's position is that 'tacit collusion is not an unconscious state'[24] While correct, the line between permissible and prohibited behavior is much more blurred under the Posner proposal.

Another issue in regard to collusive arrangements is the role of 'facilitating practices'. That term includes practices which promote collusive outcomes in the absence of explicit agreements. To the extent that this conduct is employed effectively, it substitutes for direct contact among rivals. In this area, policy actions have greater prospects for success since they focus directly on particular conduct.[25]

Among the practices that have been considered in this context are various forms of data dissemination and information exchange,[26] and also pricing structures such as the use of established basing points[27] and most favored buyer or seller provisions.[28] What these practices have in common is their promotion of interdependent price setting; which is the hallmark of tacit collusion. However, in each situation, the practice

---

[24] *Ibid.*, p. 74.

[25] For a review of facilitating practices, see Steven C. Salop, (1986) 'Practices that (Credibly) Facilitate Oligopoly Co-ordination,' J. Stiglitz and R. G. Mathewson, (eds), *New Developments in the Analysis of Market Structure*, MIT Press, pp. 265–290.

[26] See George W. Stocking, (January 1954) 'The Rule of Reason, Workable Competition, and the Legality of Trade Association Activities,' *University of Chicago Law Review*, Vol. 21, pp. 527–619; and A. M. Spence, (August 1978) 'Tacit Coordination and Imperfect Information,' *Canadian Journal of Economics*, pp. 490–505.

[27] See Carl Kaysen, (1949) 'Basing Point Pricing and Public Policy,' *Quarterly Journal of Economics*, Vol. XLIII, pp. 289–314; and Dennis W. Carlton, (April 1983) 'A Reexamination of Delivered Pricing Systems,' *Journal of Law and Economics*, Vol. XXVI, pp. 51–70.

[28] See T. C. Cooper, (Autumn 1986) 'Most-Favored Customer Pricing and Tacit Collusion,' *Rand Journal of Economics*, Vol. 17, pp. 377–388; and C. A Holt and D. T. Scheffman, (Summer 1987) 'Facilitating Practices: the Effects of Advance Notice and Best-Price Policies,' *Rand Journal of Economics*, Vol. 18, pp. 187–197.

can sometimes be used to achieve more efficient outcomes, whether for the firm or the overall market. As a result, these practices are uniformly examined under the rule of reason.

## 4. THE EXERCISE AND ABUSE OF MONOPOLY POWER

This second leg of American antitrust policy stems from the Sherman Act prohibition of 'monopolization.' Since the statutory language refers to monopolizing conduct rather than the mere presence of monopoly, there is the inevitable question of whether the conduct could exist without the result, and if not, whether it is structural monopoly that is actually prohibited.

There are various answers to this question. On the one hand is the antipathy towards dominant firms in any relevant market. For such firms, standards of proscribed conduct are particularly stringent. Still, the antitrust laws prohibit monopolizing conduct even by firms without dominant market shares. The conduct proscribed are actions designed specifically to place rivals in a disadvantaged position. Of course, it is often difficult to distinguish such conduct from competitive behavior. Predatory conduct, however, violates the antitrust laws only in the presence of market power, so that the same preliminary tests are required here as in Section 1 cases.

Policy standards against dominant firms were established originally in two early decisions, where both the Standard Oil Company and the American Tobacco Company were found guilty under Section 2 of the Sherman Act.[29] Both firms had achieved very large market shares through actions designed to exclude rivals. However, those decisions left unanswered the questions of what market shares were needed to demonstrate dominance and what were the limits of exclusionary conduct.

The classic answers to these questions appear in Judge Learned Hand's famous decision in the *Alcoa* case.[30] Hand wrote that a 90 percent market share 'is enough to constitute a monopoly; it is

---

[29] *Standard Oil Co. of New Jersey v. United States*, 221 US 1 (1911); and *United States v. American Tobacco Co.*, 221 US 106 (1911).

[30] *United States v. Aluminum Company of America*, 148 F. 2d 416 (2d Cir. 1945).

doubtful whether 60 or 64 percent would be enough; and certainly 33 percent is not.' But which percentage applied in that case depended on the appropriate definition of the market; whether the market is appropriately defined by production or sales, and whether secondary sources of aluminum ingot should be included. The sticking point there, as in many antitrust cases, was how to define the market.

The *Alcoa* decision offered an even sharper message in terms of the conduct standard to be applied. With the simple statement that 'no monopolist monopolizes unconscious of what he is doing,' the judge seemed to suggest that dominance was sufficient. However, Hand also asked whether Alcoa had purposely achieved its monopoly position or rather had monopoly thrust upon it. He indicated that a monopoly position would not violate the antitrust laws if it was justified by economies of scale, by changes in tastes, or if higher costs had driven all firms but one from the market. Thus, a dominant firm is not guilty of monopolization if it survived by 'virtue of his superior skill, foresight and industry.'[31]

As reaffirmed in the 1966 *Grinnell* decision, 'the offense of monopoly ... [requires] the possession of monopoly power in the relevant market and ... the willful acquisition or maintenance of that power as distinguished from growth or development as a consequence of a superior product, business acumen, or historic accident.'[32] So long as the defendant has gained a substantial market share through purposeful actions, that is sufficient for illegal monopolization.

Although the *Alcoa* and *Grinnell* opinions emphasize the structural or market share requirements for illegal monopolization, more recent decisions have emphasized the behavioral requirements for Section 2 violations. Thus in the *Berkey Photo* decision of 1979, the judges ruled that dominant firms 'must refrain ... from conduct directed at smothering competition,' and also may not use the 'monopoly power attained in one market to gain a competitive advantage in another.'[33] Still, in that case, the court ruled in favor of a dominant firm by finding that 'an integrated business [does not] offend the Sherman Act whenever one of its departments benefits from an association with a division

---

[31] 148 F. 2d at 430.
[32] *United States v. Grinnell Corp.*, 384 US 563 (1966).
[33] *Berkey Photo, Inc. v. Eastman Kodak Company*, 603 F. 2d 263, 275, 276 (2d Cir. 1979); cert. denied, 444 US 1093 (1980).

possessing a monopoly in its own market.'³⁴ There is a conduct component in cases charging illegal monopolization, but there is only a hazy line to judge what actions trigger this offense.

However, antitrust cases have two phases: Remedy as well as Liability, and government suits on monopolization have more frequently foundered on the former than the latter. In many such cases, government policy-makers have been uncertain as to just what remedies should be sought. Despite the successful dissolution of the American Telephone and Telegraph Company, that solution remains unpopular and there is great hesitancy to recommend it.³⁵

In two other cases, the FTC argued that more competitive structures in the consumer goods sector would result from the compulsory licensing of certain trademarks. The Commission proposed that remedy in both the *Realemon* and the Cereals monopolization cases,³⁶ but great controversy followed and neither case was successfully completed.³⁷ The problem of how to fashion an effective remedy for monopolization, short of dissolution and divestiture, remains a conundrum for this section of the law.

Even apart from the problem of Remedy, there remains a significant ambiguity in the law of monopolization. On the one hand are the market share percentages laid down in the *Alcoa* decision that even a 60 to 64 percent share is a doubtful indicator of monopoly power; while on the other is the explicit definition of monopoly power as 'the power to control prices or exclude competition.'³⁸ The latter definition suggests that market or monopoly power is achieved whenever firms have meaningful control over price, or in other words, when firm demand curves are downward sloping. In that context, market power is achieved whenever the Lerner Index exceeds zero.³⁹

---

[34] *Ibid.*, p. 276.

[35] See particulary *United States v. United Shoe Machinery Corp.*, 110 F. Supp. 295 (D. Mass. 1953), affirmed per curiam, 347 US 521 (1954).

[36] *In the Matter of Borden, Inc.*, F.T.C. Docket No. 8978, (1976), (1978); and *In the Matter of Kellog Co. et al.*, F.T.C. Docket No. 8883, (1972).

[37] See Richard Schmalensee (April 1979) 'On the Use of Economic Models in Antitrust: The *Realemon* Case, '*University of Pennsylvania Law Review*, Vol. 127, pp. 994–1050; and F. M. Scherer, (1979) 'The Welfare Economics of Product Variety: An application to the Ready-to-Eat Cereals Industry,' *Journal of Industrial Economics*, Vol. 28.

[38] *United States v. E. I. Dupont and Co.*, 351 US 377, 391.

[39] See Abba Lerner, (June 1934) 'The Concept of Monopoly and the Measurement of

This ambiguity arises from recognition that market power, in terms of control over price, or a positive value of the Lerner Index, does not require the high market shares indicated in Judge Hand's opinion. Concentrated industries, where no single firm has these shares, or firms that sell highly differentiated products which are not highly interchangeable with rival products, are two situations where market power may be present even in the absence of a dominant firm.

A possible way to reconcile these two criteria of monopoly power is to relate them to the statutory language of Section 2. That section condemns both monopolizing conduct and also attempts to monopolize. In 1946, the Supreme Court offered the following interpretation: 'attempt to monopolize' means the employment of methods, means and practices which would, if successful, accomplish monopolization and which, though fallen short, nevertheless approach so close as to create a *dangerous probability of it.*'[40]

To succeed at efforts to monopolize, a firm must already have some degree of monopoly power, for success is hardly likely with highly competitive markets. But what is relevant here is power over price, and not necessarily the dominant market shares suggested by Judge Hand. Such cases require both the presence of monopoly power and conduct designed expressly to exclude rivals from the market.[41] Private plaintiffs rather than government prosecutors frequently bring such actions where the gains from recouping treble damages provide sufficient incentive.

There are two approaches to this problem. The first follows from the recent discussion of Raising Rival's Costs, that refers to circumstances where a firm takes actions which impose higher costs on its rivals.[42] Such actions often concern relationships with dealers or suppliers, but this need not be so. For example, Williamson's discussion of the

---

Monopoly Power,' *Review of Economic Studies*, Vol. 1, pp. 157-175; and William M. Landes and Richard A. Posner,(March 1981) 'Market Power in Antitrust Cases,' *Harvard Law Review*, Vol. 94, pp. 937-996.

[40] *American Tobacco Co. v. United States*, 328 US 781, 785 (1946) italics added.

[41] See Roger D. Blair and David L. Kasserman, (1985) *Antitrust Economics*, Richard D. Irwin; Homewood, Illinois, pp. 119-121.

[42] See Steven C. Salop and David T. Scheffman, (May 1983) 'Raising Rivals' Costs,' *American Economic Review*, Vol. 23, pp. 267-271; and Thomas G. Krattenmaker and Steven C. Salop, (December 1986) 'Anticompetitive Exclusion: Raising Rivals' Costs To Achieve Power Over Price,' *Yale Law Journal*, Vol. 96, pp. 209-293.

*Pennington* case[43] describes circumstances where wage increases negotiated with an industry-wide union had differential effects on large and small firms.[44] As a result, there was the prospect for anticompetitive results.

A second approach deals with the more traditional concept of predatory conduct, over which there has been substantial controversy in recent years. On the one hand is the view that predatory conduct is unlikely because it requires a commitment to conduct not consistent with profit-maximization.[45] Even if irrational, some firms may attempt such actions, which led to the antitrust rule proposed by Areeda and Turner.[46] They recommend that predatory pricing be defined as prices less than marginal costs when average costs are declining, or average costs when these costs are constant or increasing. In either case, setting price below these benchmarks does not maximize short-run profits and is therefore narrowly irrational. In this sense, the Areeda-Turner rule is consistent with McGee's earlier position.

The problem with this approach is its attempt to explain predatory conduct in a non-strategic context. Yet predatory conduct is essentially strategic in that it is designed to force certain types of conduct or actions on rival firms.[47] The growing recognition that predatory conduct rests on strategic foundations requires a renewed emphasis on the purpose or intent of particular conduct before determining its antitrust consequences.

The most recent Supreme Court decision in this area is the *Matsushita* decision of 1986.[48] In that case, the Court majority may

---

[43] *United Mine Workers v. Pennington*, 85 S. Ct., 1985 (1965).

[44] Oliver E. Williamson, (February 1968) 'Wage Rates as a Barrier to Entry: The Pennington Case in Perspective,' *Quarterly Journal of Economics*, Vol LXXXII, pp. 85–116.

[45] See John S. McGee, (October 1958) 'Predatory Price Cutting: The Standard Oil (N.J.) Case,' *Journal of Law and Economic*, Vol. 1, pp. 137–169.

[46] Philip Areeda and Donald F. Turner, (Feb. 1975) 'Predatory Pricing and Related Practices Under Section II of the Sherman Act,' *Harvard Law Review*, Vol. 88, pp. 697–773.

[47] For some studies of predatory conduct in this vein, see F. M. Scherer, (March 1976) 'Predatory Pricing and the Sherman Act: A Comment,' *Harvard Law Review*, Vol. 89, pp. 869–890; Oliver E. Williamson, (December 1977) 'Predatory Pricing: A Strategic and Welface Analysis,' *Yale Law Journal*, Vol. 87, pp. 284–340; and William S. Comanor and H. E. Frech III, (May 1984) 'Strategic Behavior and Antitrust Analysis,' *American Economic Review*, Vol. 74, pp. 372–376.

[48] *Matsushita Electric Industrial Company v. Zenith Radio Corp.*, 475 US 574 (1986).

have accepted the skeptic's view of this issue in its comment 'that predatory pricing schemes are rarely tried, and even more rarely successful.'[49] However, the case dealt with an alleged predatory pricing scheme by a group of firms rather than a single dominant firm, which the Court suggested, was a less likely event. Moreover, the decision stated explicitly that it did not seek to resolve the debate over the appropriate criterion for predatory pricing.[50] That result must await another case.

## 5. MERGERS BETWEEN RIVALS

Enforcing the Clayton Act provisions against anticompetitive mergers is the joint responsibility of the FTC and the Department of Justice. As a result, one of the two agencies takes responsibility for evaluating specific acquisitions. To facilitate the process, Section 7A of the Clayton Act, enacted in 1976 as the Hart-Scott-Rodino Act, requires advance notification and the provision of certain data for large mergers.

In this policy area, government actions are particularly critical because few private parties have both the incentive and standing to intervene. While customers, including final consumers, are harmed by acquisitions leading to higher prices and lower outputs, this result takes place in the future so there can be no current damages to trigger a lawsuit. On the other hand, rivals benefit when higher prices are charged, so often there is no incentive for them to challenge a merger. Only when rivals fear that an acquisition will limit their own position in the market will they wish to prevent the merger, but here they first must demonstrate that such results flow directly from anticompetitive acts and not merely from more efficient operations. Both considerations limit private enforcement of the antimerger provisions.[51]

The basic premise of this policy is that the structure of the market matters. What this means is that market results are affected by the number and relative sizes of rival firms. Although some have argued against this presumption, it remains the basic premise of antitrust

---

[49] *Ibid.*
[50] *Ibid.*, p. 1355 n. 8.
[51] *Cargill v. Monfort*, 107 S. Ct. 484, 492-493 (1986).

actions against corporate mergers. The evidence supporting this premise is largely the extensive empirical literature that relates profitability to various dimensions of market structure. There is also a more limited empirical literature dealing with actual prices. Many of these studies are reviewed elsewhere.[52]

There is also evidence from past antimerger cases. In one such study, the authors review an action brought by the Federal Trade Commission in 1981 against Xidex Corporation, the major producer of duplicating microfilm.[53] Consistent with the cross-section evidence, the authors find that the merger of two leading firms led to substantial price increases. Furthermore, the higher prices yielded sufficiently greater profits to permit the firm to recover its costs of the acquisition in about two years.

In another study, Elzinga examines the effectiveness of relief decrees in Government antimerger cases through 1960.[54] Since government prosecutors were apparently successful in demonstrating the anticompetitive effects of these mergers, he considers whether acquired firms were eventually divested and if the reconstituted firm was both independent and viable. Of the 39 cases in his sample, Elzinga reports that 29 relief orders were either 'deficient' or represented 'unsuccessful relief.' His conclusion is that the merging firms retained the ability to set high prices and reap high returns; and therefore he questions whether past antimerger efforts have been very effective.

Even where mergers lead to higher prices, the policy implications of this result have been questioned. In a 1968 article, Williamson compares the efficiency loss from the increased prices presumed to follow from horizontal mergers with efficiency gains that might also result. Such gains could occur from either the more rapid realization of scale economies or simply from permitting more efficient firms to acquire their high-cost rivals. By comparing measures of consumer welfare

---

[52] Harold Demsetz, (1974) 'Two Systems of Belief About Monopoly' and Leonard W. Weiss, (1974) 'The Concentration-Profits Relationship and Antitrust,' Harvey J. Goldschmid *et al.*, *Industrial Concentration: The New Learning,* Boston: Little Brown, pp. 164-232.

[53] David M. Barton and Roger Sherman (December 1984) 'The Price and Profit Effects of a Horizontal Merger: A Case Study,' *Journal of Industrial Economics*, Vol. XXXIII, pp. 165-177.

[54] Kenneth Elzinga, (April 1969) 'The Anti-merger Law: Pyrrhic Victories?' *Journal of Law and Economics*, Vol. 12, pp. 43-78.

from lower costs with those resulting from higher prices, Williamson finds that price effects must be large to offset the effects of any lower costs that result from a merger.[55] He concludes that the price-enhancing effects of a merger must be relatively large to offset any cost savings that may also result.

How to account for any lower costs from a merger is subject to some dispute. While the basic form of the trade-off is apparent, the policy implications remain uncertain because the prospective cost savings could arise even in the absence of the merger. If economies of scale are present, so the counter-argument proceeds, these gains can be realized through internal expansion so that no merger is required. The harmful price effects of a merger need not occur even when the cost savings are achieved. Internal expansion acts as a safety valve that permits a stronger policy against mergers.

This issue has never been fully resolved. Whether the presence of cost-saving efficiencies is sufficient to validate an otherwise anticompetitive merger depends on the time periods involved as well as the relative magnitudes. How much sooner must cost-savings be achieved to obviate the long-standing effects of higher prices? While formal answers for this question can be provided, they offer few insights for effective policy-making. And the current policy position is that economies are merely one factor to be considered rather than providing full justification for an otherwise anticompetitive merger.

In this policy area, unlike the previous two, the Antitrust Division has issued various guidelines. The first set of merger guidelines was announced by Donald F. Turner, then head of the Antitrust Division, in May 1968. Although the Division acknowledged that it had no control over court decisions, so that actual judicial results might differ from the proposed standards, this statement set the tone of antimerger policy for a considerable period. Fourteen years later, on June 14, 1982, the Antitrust Division, then headed by William F. Baxter, adopted a new set of guidelines. And two years later, on June 14, 1984, further guidelines were announced which extended some of the principles announced two years earlier.

These statements provide explicit policy standards, and it is

---

[55] Oliver E. Williamson, (March 1968) 'Economies as an Antitrust Defense,' *American Economic Review*, Vol. 58, pp. 18–36.

interesting to consider how the more recent guidelines differ from those published in 1968; and also how they differ from the existing pattern of judicial decision. The short answer to these questions is that while there are major differences, there are important similarities as well.

Even under the more recent guidelines, the basic principles for antimerger enforcement practices remain those established in the Supreme Court decision in the *Philadelphia National Bank* case.[56] The approach taken there is first to define a relevant market and then determine the competitive effects of a merger largely from the resulting changes in market shares. While other factors are considered, evidence on market shares, measured within a relevant economic market, offers the clearest guide to the competitive implications of a particular merger.[57]

A striking feature of the more recent guidelines is their explicit acceptance of the position that concentration and market shares affect prospects for collusion. Thus, the recent guidelines observe: 'where only a few firms account for most of the sales of a product, those firms can in some circumstances coordinate ... their actions ... to approximate the performance of a monopolist.[58] This passage represents a direct rejection of the view that market structure is unimportant in describing competitive effects.

Despite the similarities between the 1968 and the more recent guidelines, there are important differences as well. Of particular importance is the added discussion of market definition. Any evaluation of market shares is only as good as the underlying definition, so the more recent statements are more precise in regard to the criteria to be used.

The economic construct of a market rests on substitutability among products in both production and consumption, as reflected in crosselasticities of supply and demand. However, rather than evaluating these parameters by looking directly at the chain of substitutes, the

---

[56] *United States v. Philadelphia National Bank*, 374 US 321 (1963).

[57] While the recent guidelines state that easy entry conditions are relevant to the competitive effects of a merger, the Supreme Court has never held that this factor rebuts a showing of high market shares under the *Philadelphia National Bank* standards. However, a Court of Appeals decision has ruled that easy entry validates a merger even when the combined market show was nearly 50%. See *United States v. Waste Management, Inc.* 743 F. 2nd 976 (1984).

[58] US Department of Justice, (June 14, 1982) *Merger Guidelines*, duplicated, p. 2.

more recent guidelines deal instead with the ability of a hypothetical monopolist, consisting of all firms in a proposed market, to raise price effectively. Thus, 'a market is defined as a product or group of products or geographic area in which it is sold such that a hypothetical, profit-maximizing firm not subject to price regulation, that was the only present and future seller of those products in that area would impose a "small but significant and non-transitory" increase in price above prevailing or likely future levels.'[59]

While this definition emphasizes pricing constraints rather than substitutes, it necessarily rests on the relative degree of substitutability of both supply and demand. The hypothetical monopolist would only be effective in its price-raising efforts if the relevant degrees of substitutability were sufficiently low, as reflected in the appropriate cross-elasticities of supply and demand. To acknowledge this fact, the authors state in the very next paragraph that 'in determining whether one or more firms would be in a position to exercise market power, it is necessary to evaluate both the probable demand responses of consumers and the probable supply responses of other firms.[60] The paragraph goes on to describe various types of demand and supply responses. The task of defining markets invariably returns to questions of substitutability.[61]

Note that the issue of market definition could be avoided by describing directly the effect of a merger on the increased ability of firms to raise prices. An alternate approach, therefore, would be to evaluate the competitive effects of a merger directly without reference to a relevant economic market and the market shares involved. But that approach would reject the methodology of the *Philadelphia National Bank* case and was not adopted in the new guidelines. Instead, the more recent statements pay greater attention to how markets should be defined.

A second consideration which distinguishes the more recent guidelines from the earlier version is the market share test to be applied.

---

[59] US Department of Justice, (June 14, 1984) *Merger Guidelines*, duplicated, paragraph 2.0.
[60] *Ibid.*
[61] Some further discussion of these issues appears in Janusz A. Ordover and Robert D. Wilig, (March 1983) 'The 1982 Department of Justice Merger Guidelines: An Economic Assessment,' *California Law Review*, Vol. 71, pp. 535-574.

While both the 1968 guidelines and the courts had considered the market shares of the merging firms and the concentration ratio of the entire industry, the more recent statement applies the Herfindahl-Hirschman Index (HHI) of market concentration.[62] In contrast to the four-firm concentration ratio, this index is affected by the distribution of market shares among the top firms. The index is higher with unequal market shares and lower with greater equality, a factor not captured before. And the HHI is particularly sensitive to the market shares of the leading two firms.

Economists have long acknowledged that the Herfindahl-Hirschman Index has theoretical advantages. Yet this measure and the four-firm concentration ratio are highly correlated; Scherer reports that the correlation coefficient across industries between the two measures is 0.94.[63] Moreover, both perform equally well in most empirical studies of the effect of market structure on various dimensions of performance.[64] Thus, it is questionable whether the new measure adds much to decisions taken on individual cases.[65]

The 1984 guidelines gave additional emphasis to foreign competition. The statement makes clear that the market shares of foreign suppliers need also be considered in assessing the competitive implications of particular mergers. If increased imports result from domestic price increases, foreign suppliers should also be included in the relevant market.

The guideline's purpose was to provide benchmarks to decide when a merger will or will not be challenged. The stated thresholds, which are the same in the 1982 and 1984 guidelines, are not so different from those used in the past. For values of the HHI below 1000, markets are considered unconcentrated and few mergers will be challenged. An index of 1000 is equivalent to ten equal-sized firms, which suggests a four-firm concentration ratio exceeding 40 percent.

Between 1000 and 1800, markets are moderately concentrated. The

---

[62] This index is the sum of the squares of market shares for all firms in the market, experssed as percentages rather than proportions.

[63] F. M. Scherer, (1980) *Industrial Market Structure and Economic Performance*, second edition, Chicago: Rand McNally, p. 58.

[64] *Ibid.*, pp. 279-280.

[65] This point is acknowledged by the author of the revised guidelines. See William F. Baxter, (March 1983) 'Responding to the Reaction: The Draftman's View,' *California Law Review*, Vol. 71, pp. 625-626.

upper limit is equivalent to five and one-half or six equal-sized firms, which corresponds roughly to a concentration ratio of 70 percent. The Department of Justice noted that it would challenge mergers in such markets if they raise the index by 100 points or more. An increase of this magnitude would occur if the product of the market shares of the two merging firms equals 50, such as shares of 2 percent and 25 percent, or 5 percent and 10 percent.

In markets with values exceeding 1800, the guidelines state that mergers will generally be challenged if the index is increased by even 25 points. That level would be reached, say, by a merger between firms with market shares of 4 percent and 6 percent, respectively.

The Department of Justice completed a statistical study of market structures in the US economy. It reported that an index of 1000 corresponds empirically to a four-firm concentration ratio of approximately 50 percent, and a value of 1800 to a ratio of about 70 percent. These values are very close to those noted above.

It is striking that similar standards have long been used to evaluate the competitive effects of horizontal mergers. Indeed, the 1968 guidelines suggest a benchmark of 75 percent for the four-firm concentration ratio. With a single exception, the major merger cases of the past would still be challenged even if the new guidelines had been applied.[66]

In one specific area, the 1984 guidelines departed from the version published two years earlier. A section was added on efficiencies which stated that these would be considered before deciding whether to challenge a merger.[67] However, the Attorney General also announced 'that efficiencies do not constitute a defense to an otherwise anticompetitive merger'.[68] While another factor was added to the calculus, it was not accorded exclusive significance.

However, despite the basic similarity of the written standards, the actual pace of enforcement activity has slackened considerably. Many fewer cases have been brought, and as a result, the pace of merger

---

[66] Donald I. Baker and William Blumenthal, (March 1983) 'The 1982 Guidelines and Preexisting Law,' *California Law Review*, Vol. 71, p. 334.

[67] US Department of Justice, (June 14, 1984) *Merger Guidelines*, duplicated, paragraph 3.5.

[68] Attorney General William French Smith, Press Release, US Department of Justice, June 14, 1984, p. 15.

activity has expanded considerably.[69] There is much doubt, therefore, as to whether current enforcement officials have followed their own standards.

## 6. MERGERS BETWEEN BUYERS AND SELLERS

The evaluation of vertical mergers has changed considerably in recent years. Originally, vertical mergers were considered and decided under a theory of foreclosure.[70] That theory describes the prospective effect on rivals at a particular stage of production from creating a newly integrated firm. There is an implicit assumption that each merging party deals only with his intra-firm supplier or purchaser so that independent, single-stage rivals are effectively foreclosed from the business. The latter can then be 'squeezed' from the market place which leads the newly integrated firm's market share to rise in one of the vertically connected markets. For example, if a firm with a 50 percent market share at an early stage acquires another with 10 percent share at the only succeeding stage, then on account of the merger, rivals of the latter firm are said to be foreclosed potentially from half the market for needed supplies.[71]

The *Brown Shoe* decision provides a clear statement of this theory. 'The primary vice of a vertical merger or other arrangement tying a customer to a supplier is that, by foreclosing the competitors of either party from a segment otherwise open to them, the arrangement may act as a 'clog on competition' which 'deprives rivals of a fair opportunity to compete.' Every extended vertical arrangement by its very nature, for at least a time, denies to competitors of the supplier the opportunity to compete for part or all of the trade of the customer party to the

---

[69] Between 1982 and 1987, the enforcement rate by the FTC and the Department of Justice was 0.7% on all reported mergers, as compared to 2.5% in 1979 and 1980, prior to the Reagan Administration. During that same period, the number of premerger transactions reported increased from 868 in 1979 and 824 in 1980 to 2406 in 1986 and 2254 in 1987. Peter Rodino, Chairman, Committee on the Judiciary, US House of Representatives, News Release, March 2, 1988, p. 3.
[70] Philip Areeda and Donald F. Turner, (1980) *Antitrust Law* Vol. IV Boston: Little Brown, pp. 296–319.
[71] See William S. Comanor, (May 1967) 'Vertical Mergers, Market Power, and the Antitrust Laws,' *American Economic Review*, Vol. LVII, pp. 254–265.

vertical arrangement.' The decision goes on to indicate that the anticompetitive effect of a vertical merger depend on 'the size of the share of the market foreclosed.[72]

The 1968 merger guidelines rest on the same theory of foreclosure by emphasizing the market shares of the merging firms. So long as the supplying firm accounts for a least 10 percent of sales in its market, and the buying firm has over 6 percent of purchases, the merger would be challenged unless it could be demonstrated that there were no significant entry barriers into the relevant markets. While these percentages gave explicit meaning to the principles contained in the *Brown Shoe* decision, they offered no additional insight into the competitive effects of these mergers.

While the theory of foreclosure may describe the process by which one firm extends its market position from one vertical stage to another, that extension does not necessarily imply an expansion of market power. The maximum profits that could be earned from a well-defined vertical stream of production is that which could be earned by a fully integrated monopolist. However, a single-stage monopolist can achieve the same returns. When that is the case, the extension of a firm's market position to preceding or succeeding stages does not increase its profits nor raise final consumer prices. The extension of market position does not necessarily contribute to an increase in market power.

However, there may be circumstances where foreclosure can extend market power. This process can limit the ability of single-stage rivals to compete, or even to remain in the market, so that dual-stage entry becomes essential. And where dual-stage entry is more difficult, for one reason or another, there may be greater entry restrictions and an increase in market power. What is required is to demonstrate the particular circumstances through which a particular vertical merger accentuates existing degrees of market power. And that explanation is not provided by the traditional theory of foreclosure.

Accordingly, both the 1982 and 1984 merger guidelines suggest a different approach. The statements offer three theories under which vertical mergers would be challenged. The first and most important concerns the possible effect of such mergers on restricted conditions of

---

[72] *Brown Shoe Co. v. United States*, 370, US 323, 324 (1962).

entry. Three conditions are required: a) the merger must lead to the requirement of dual-stage entry; b) simultaneous entry into both markets must be more difficult and costly than single-stage entry; and c) markets must already be highly concentrated. Note that the first of these conditions is really a new packaging of the old theory of foreclosure.

There are two other theories under which vertical mergers can be anticompetitive. The first exists where a vertical merger facilitates collusion at the original stage of production. The authors suggest that sometimes, retail prices may be more easily monitored than those at the preceding stage so that a vertical merger would promote collusive outcomes. And the second deals with instances where a vertical merger is used to avoid government rate regulation.

While the more recent guidelines represent a greater departure from the past in this area than in regard to horizontal mergers, they create a more coherent set of policy standards. To this point, however, there have been few such cases brought under the new guidelines so that a more certain judgement must wait for the future.

## 7. VERTICAL CONTRACTUAL RELATIONSHIPS

While vertical arrangements refer to all contractual relationships between buyers and sellers, antitrust enforcement has emphasized two types of circumstances. The first category includes both resale price maintenance and customer and territorial restraints, agreements that limit competition among distributors and promote higher distribution margins. The second includes exclusive-dealing arrangements which restricts the freedom of buyers to make purchases from other suppliers. Both types of restraints are enforced by the threat that the seller or manufacturer will no longer supply the particular distributor, and both have been subject to much antitrust scrutiny. At this point, we consider the first category of these restraints.

The antitrust treatment of vertically-imposed price restraints has vacillated for over half a century. Judgements have fluctuated from antagonism to hostility, then to antagonism once again, and most recently to a growing approval. The milestones of this path are readily traced. As far back as 1911, the Supreme Court condemned vertical price fixing as a *per se* violation of the antitrust laws.[73] The court ruled that a manufacturer, 'having sold its product at prices satisfactory to itself, the public is entitled to whatever advantage may be derived from competition in the subsequent traffic.'[74]

By the 1920s, business views were ascendant and the interests of manufacturers given greater credence. And such views were strengthened by protectionist attitudes generated by the Great Depression. In this setting, resale price maintenance, another term for vertical price fixing, was specifically exempted from Sherman Act prohibitions so long as the products were trademarked or branded and in 'free and open competition with commodities of the same general class.' This exemption was strengthened by the McGuire Act of 1952. Throughout this era, and into the 1970s, vertical price fixing remained an approved form of business conduct.

In the 1970s, the policy reversed itself once again. Both earlier statutory provisions were repealed and the original antitrust prohibitions contained in *Dr. Miles* permitted to remain. The fear that competitive prices would drive large numbers of distributors out of business had dissipated. What was stressed instead was that consumers gain from lower prices and that the restraints prevented prices from being as low as possible.[75]

The antitrust treatment of vertical non-price restraints is more recent in origin but no less varied. These restraints typically concern the territories within which distributors might sell or the types of customers to whom they can sell. Despite the general similarity in their economic consequences, the antitrust response has been quite different. The first case to reach the Supreme Court on these restraints was only in 1963,[76] where the Court adopted an agnostic position.

Four years later, however, the Court moved sharply towards treating both price and non-price restraints the same. In *Schwinn*, it ruled that so long as 'all indicia of ownership' had been transferred to the original buyer, territorial and customer restrictions were *per se* violations of the Sherman Act.[77] After an original hesitation, a consistent standard was

---

[73] *Dr. Miles Medical Co. v. John D. Park & Sons Co.*, 220 US 373 (1911).
[74] 220 US at 409.
[75] Richard I. Gibbens and Laura P. Worsinger, (1977) 'Vertical Restraints After Repeal of Fair Trade,' *Fordham Law Review*, Vol. 45, pp. 671–675.
[76] *White Motor Co. v. United States*, 372 US 253, 263 (1963).
[77] *United States v. Arnold Schwinn and Company*, 388 US 365 (1967).

set. Together with the 1975 legislation, that decision represented the high-water mark of antitrust rejection of vertically imposed restraints.

Since that time, the law has been in general retreat; it occurred first with non-price restraints. The case involved the distribution of television sets and the Supreme Court ruled that vertical restraints may 'promote inter-brand competition by allowing the manufacturer to achieve certain efficiencies in the distribution of his products' so that 'the *per se* rule stated in *Schwinn* must be overruled.'[78] In effect, the Court agreed that the manufacturer's interest in securing an efficient distribution network by protecting distribution margins coincides with the interest of consumers. Manufacturers were free to determine their distribution arrangements.

Justice White, in his concurring opinion, noted the full implications of *Sylvania* for vertical restraints. He observed that the same economic effects achieved through non-price restraints can be fashioned more directly through setting resale prices. Therefore, 'the effect, if not the intention, of the Court's opinion is necessarily to call into question the firmly established *per se* rule against price restraints.[79]

Following Justice White's comment, it was only a matter of time before the Supreme Court faced the inconsistency of existing policy standards towards vertical price and non-price restraints. In 1982, the issue was raised directly in a brief filed by the Department of Justice.[80] The brief argued that 'in many cases, resale price maintenance may have the same effect as the non-price measures Sylvania removed from the category of *per se* offenses: they may be highly pro-competitive and enhance consumer welfare by stimulating inter-brand rivalry.' For this reason, the Department continued 'the logic of *Sylvania* compels the conclusion that resale price maintenance—like other vertical restrictions—is unsuitable for *per se* treatment.'[81]

Despite this argument, the Supreme Court refused to consider the merits of the issue and allowed the *per se* rule to remain.[82] However, the scope of the prohibition has been progressively narrowed. As indicated in the *Sharp Electronics* decision, the goal of minimum consumer

---

[78] *Continental T.V. v. Sylvania*, 433 US 36 (1977).
[79] 433 US 36 at 69–70 (1977).
[80] Brief for the United States as *amicus curiae* in support of Petitioner, *Monsanto Co. v. Spray-Rite Corp.*, No. 82-914, p. 6.
[81] *Ibid.*, p. 19.
[82] *Monsanto Co. v. Spray-Rite Corp.*, 104 US 1464, (1984).

prices is no longer paramount.[83] In that case, the Supreme Court ruled that unless there was a definite agreement over specific prices, *per se* treatment was not warranted. It was no longer sufficient that a manufacturer discontinued sales to a low-priced distributor because of objections raised by a higher-priced rival.[84]

For the most part, the shifting legal treatment of these restraints has followed the economic arguments, although with a lag, Telser's 1960 article was influential in explaining why manufacturers might wish to impose these restraints and Bork's 1966 paper added an efficiency rationale.[85] So long as output is increased, as was part of Telser's explanation, then economic efficiency is promoted and the interest of manufacturers and consumers are joined. Indeed, the current position of the enforcement agencies is precisely that.[86]

However, the normative economics of vertical restraints is not so simply stated. Bork's conclusions rest on the implicit assumption that all consumers are alike in terms of their preferences for the distribution 'services' supported by these restraints. On the other hand, when differences arise, and particularly when those consumers who value the product most have the least need for the distribution 'services,' then the opposite conclusion may follow. While manufacturers may still gain from the imposition of these restraints, consumers may be worse

---

[83] *Business Electronics Corp. v. Sharp Electronics Corp.*, 88 Daily Journal D.A.R. 5504, 5507 (1988).

[84] *Ibid.*, p. 5505.

[85] Lester Telser, (1960) 'Why should manufacturers want Fair Trade?' *Journal of Law and Economics*, Vol. 3, p. 3; Robert H. Bork, (1966) 'The Rule of Reason and the Per Se Concept: Price Fixing and Market Division,' *Yale Law Journal*, Vol. 25, p. 373. Those papers have been followed by a long and extensive literature on this and related topics. They are reviewed in William S. Comanor, 'The Two Economics of Vertical Restraints,' *Review of Industrial Organization*, forthcoming.

[86] Since 1981, the Department of Justice has brought no new cases that challenge vertical restraints, and the FTC has issued only four complaints, 'all before Reagan appointees attained a majority.' See Alan Fisher *et at.*, (Fall 1987) 'Do the DOJ Vertical Restraints Guidelines provide guidance?' *Antitrust Bulletin*, Vol. XXXII, p. 634. Furthermore, in January 1985, the Department of Justice published its *Vertical Rstraints Guidelines* that reflected its view that such restraints generally promote economic efficiency. However, unlike the earlier *Merger Guidelines*, these 'have had virtually no impact on the legal and judicial community,' and have been largely 'ignored.' These conclusions appear in Ernest Gellhorn and Kathryn M. Fenton, (Fall 1988) 'Vertical Restraints during the Reagan Administration, A Program in Search of a Policy,' *Antitrust Bulletin*, Vol. XXXIII, p. 567.

off.[87] The final standards towards these restraints are still to be set.

## 8. EXCLUSIVE-DEALING ARRANGEMENTS

The second category of vertical restraints deals more directly with exclusionary effects. When these restraints are imposed, distributors are required to buy from only a single producer on pain of not receiving any shipments from the seller. Similar restraints deal with the quantity or range of products to be purchased. All such requirements have the effect of reducing the range of choice available to individual buyers and preventing rival sellers from making similar transactions with these distributors.

Exclusive-dealing arrangements that tie a buyer to a particular seller are a contractual form of vertical integration. While integration is not illegal, there is wide hostility to some of its effects.[88] However, as with vertical mergers, the foreclosure theory has been criticized. Even if foreclosure precluded a firm from actively competing at a particular stage of production, this effect was limited to the particular stage and could not affect dual-stage operations. Furthermore, whatever the effects of foreclosure on market shares, there were no necessary consequences for market power.

While this critique rests on a valid theoretical structure for single-product firms, it is far less applicable to multiproduct firms where there is the prospect for substantial economies of scope. Such economies may be pervasive at the distribution stage where costs can be much lower from selling a large number of products. In such circumstance, there may be little option for rival producers to distribute their own products if a particular set of distributors is foreclosed to them. Similarly, where scale economies in distribution are considerable, the foreclosure of firms from particular markets may also have important anticompetitive results.

Anticompetitive effects are particularly likely when exclusive dealing

---

[87] William S. Comanor, (March 1985) 'Vertical Price-Fixing, Vertical Market Restrictions, and the New Antitrust Policy,' *Harvard Law Review*, Vol. 98, pp. 983–1002.
[88] See *Brown Shoe Co. v. United States*, 370 US 294 (1962).

raises rivals' distribution cost. In the face of substantial economies of scope or scale in the distribution sector, rivals' costs are increased by the differential costs with the next best distribution process, and established firms can increase their prices accordingly. These results apply to the relationship between a dominant firm and a fringe of smaller firms already in the market or with prospective entrants. In either case, vertical arrangements may be used to expand the cost differential between rivals with corresponding effects on market power.[89]

## 9. SOME CONCLUSIONS

American antitrust policy, both now and throughout the one hundred years since its inception, is not a settled doctrine but rather a continually changing set of restrictions imposed on the unregulated sector of the economy. At the current time, some provisions are affected both by economic and legal knowledge but even more so by economic conditions. As the economy changes and as different economic problems arise and recede, the law changes as well.

The antitrust laws provide a constitution for large segments of the US economy. The general goal is one of promoting competition but what this means and how it is applied has varied over the past century. And the content of these basic provisions will continue to change in the future. Yet, in its basic adherence to the objective of promoting a more competitive economy, there is a commitment to a particular economic system. And it is in terms of this commitment that the antitrust laws have had their greatest impact.

---

[89] See William S. Comanor and H. E. Frech III, (June 1985) 'The Competitive Effects of Vertical Agreements?' *American Economic Review*, Vol. 75, pp. 539–546. For a more general treatment of actions taken to raise rivals' costs, see the references in note 42.

# Canadian Competition Law: 100 Years of Experimentation

**LEONARD WAVERMAN**[1]
*University of Toronto*

## 1. INTRODUCTION AND SUMMARY

Canadian competition law predated US antitrust law. In 1889 an 'Act For The Prevention and Suppression of Combinations Formed in Restraint of Trade' was passed by the Canadian parliament.[2] To understand the developments of Canadian antitrust law in the next 100 years one must begin with certain peculiarities of the Canadian Constitution. Under the British North America Act, jurisdiction over criminal law was given to the federal government but the provinces were given jurisdiction over 'property and civil rights', i.e. civil law. While the Federal government has broad regulatory powers under a civil 'Trade and Commerce' clause, it was not until the post-Second War period that Competition Policy was interpreted as falling under this mandate.[3] Moreover, until the 1960s the highest Court of Appeal in Canada was the British Privy Council. Therefore for much of its history, federal regulation of 'unsavory' business practices was limited to criminal

---

[1] I would like to thank Gordon Kaiser, Shaym Khemani, Michael Trebilcock and Fred Webber for their insights and suggestions. Any interpretations, views and errors are mine alone.

[2] S. C. 1889 c. 41. See Bureau of Competition Policy (1989) for an excellent and concise discussion of the evolution of the Canadian Law. See Kaiser and Nielsen-Jones, 1986, for an excellent legal analysis.

[3] The courts have upheld various portions of the *Combines Investigation Act* under the 'general regulation of trade branch' of the trade and commerce power (section 91(2) of the *Constitution Act*, 1867. In 1962 Gosse wrote 'There is a good possibility that the present [combines] legislation would now be held valid under ['The Regulation of Trade and Commerce'] although forty years ago there would have been no likelihood of such a finding (p. 253).' See also footnote 4.

offenses or forays into civil law which the Privy Council considered within Canadian federal jurisdiction.[4] The cases were heard and dealt with by the normal justice system.[5]

Thus, the burden of proof for the prosecution was to prove the offence 'beyond a reasonable doubt', rather than 'on the balance of probabilities' common to civil law. Beyond a reasonable doubt has been interpreted to mean that the evidence must 'be inconsistent with any other rational conclusion . . .'.[6] This burden of proof (understandably necessary for offenses such as murder) proved difficult for offenses such as an 'anticompetitive' merger or monopoly. Additional questions of evidence (data generally, issues of circumstantial evidence) also hampered adjudication of economics-driven questions. For example, the judgement of price-fixing agreements, is based on past events while the analysis of mergers is based on prospective events and hence is difficult to prove in a criminal context.

There were no successful prosecutions of mergers in Canada and only one successful prosecution of a monopoly[7] (and that for clear exclusionary and 'criminal' activities) between 1889 and 1986 (the *Act* was recently and extensively revised in an enactment on June 19,

---

[4] In the first part of this century, attempts by the Canadian Federal government to institute civil processes failed. Two statutes enacted by the federal government in 1919, the Combines and Fair Prices Act (S.C. 1919 C. 45) and the Board of Commerce Act (S.C. 1919 C. 37) were struck down by the Judicial Committee of the Privy Council in November 1921 as *ultra vires* the federal government since they interfered with property and civil rights of the provinces. A 1935 foray of the feds into civil procedures (S.C. 1935 C. 54) was held *ultra vires* for the same reasons.

There have been numerous additional challenges to the constitutional authority of civil procedures in competition policy. In 1989, the Supreme Court of Canada approved civil procedures allowing for private damages flowing from criminal provisions of the *Act* and at the same time suggested that the entire *Competition Act* could be upheld under the Federal government's trade and commerce clause. (*General Motors of Canada Ltd. v. City National Leasing* 1989 ISCR 641, 24 C.P.R. (3d) 417.

A merger between 2 Quebec waste-rendering firms is still today before the courts as to the constitutionality of the civil merger provisions of the 1986 *Competition Act*. (Alex Couter Inc. and Senimal Industries Inc. application to the Quebec Superior Court that a June 18 1987 divestiture application by the Director to the Competition Bureau is *ultra vires*).

[5] In 1976, certain practices (refusal to deal, consignment selling, exclusive dealing, market restriction, tied selling) were made reviewable by a specialized Tribunal—the Restrictive Trade Practices Commission. A 'reviewable' offence is one where a *per se* rule of illegality does not apply.

[6] *Hodge's Case* (1838), 2 Lewin 227 at p. 228, 168 E.R. 1136.

[7] *Eddy Match Co. et al. v. The Queen* (1954), 109 C.C.C. (p. 20).

1986).[8] Canadian competition law proved to be more successful against conspiracies (price-fixing and the like) and certain practices such as misleading advertising. These practices are more akin to criminal conspiracies and fraud than are a merger or a monopoly.

The distinction between criminal and civil law was a given in the law's development. However, the confines of criminal strictures cannot alone explain the history of Canadian antitrust law, especially as applied to mergers and monopolies. It is always difficult to generalize a century of developments, but several issues are key. In a country the market size of Canada, there is always a tension between the desire to promote competition domestically and the belief that Canadian industry suffers from diseconomies of scale and is disadvantaged in world markets.[9] This tension exists because politicians and the drafters of various amending legislation (that which passed and that which too often died) have not until recently, articulated a working definition of 'competition'.[10] The courts often examined rivalry between competitors without regard to the impact on competition.[11] There has not been a clear understanding of whether antitrust law is aimed at promoting competition between domestic producers (concentration ratios are normally estimated for domestic shipments) or between all producers. Nor is there an explicit view as to the welfare function to be maximized, be it consumers' and producers' surplus or just consumers' surplus. There has not been an understanding of the relationship between Canadian protection against imports and the resulting level of domestic competition.[12]

---

[8] *Competition Act*, c. 26.

[9] In 1989, the Bureau of Competition Policy wrote, 'In many cases Canadian markets are small relative to the efficient scale of production and distribution. This imposes cost disadvantages on Canadian firms competing in international markets'. (p. 3).

[10] The 1986 Act added a preamble not present before (Section 1.1)—'The purpose of this Act is to maintain and encourage competition in Canada in order to promote the efficiency and adaptability of the Canadian economy, in order to expand opportunities for Canadian participation in world markets while at the same time recognizing the role of foreign competition in Canada, in order to ensure that small and medium-sized enterprises have an equitable opportunity to participate in the Canadian economy and in order to provide consumers with competitive prices and product choices'. The 1976 amendments to the *Combines Investigation Act* included the first statement of the role of competition.

[11] In 1956 Bladen and Stykolt wrote '. . . Canadian combines policy has paid too much attention to form and too little to effects' (p. 45).

[12] See Eastman and Stykolt (1967).

These strains can best be seen at the times when a desire is articulated to alter antitrust law. In the mid 1960s the Crown perceived a total failure of antimerger policy. The *Act* was however not substantially altered until 1986.[13] The many attempts at new legislation withered because of the inability to articulate the goals of competition policy within the reality of a small open economy. Thus no broad coalition arose which supported a new *Act*. The monopoly and merger provisions were finally altered in 1986 by a process of allowing business lobby groups (basically large firms) the ability to help formulate the new *Competition Act*, before it was presented to Parliament.[14]

The tensions between opposing views on the necessity of large firm size are evident in the merger and monopoly provisions even today. The tensions do not appear to exist when the focus is on individual firm behaviour rather than on the optimal structure of an industry. For example, practices such as vertical restraints or misleading advertising are viewed as clearly reducing welfare (which they need not do) and are dealt with accordingly.

There is an unfortunate dichotomy in the Canadian law between issues related to firm size and industry structure on one hand and optimal behaviour on the other hand. As a result the Canadian law carefully (to some too carefully) examines the virtues of large scale in merger and monopoly cases while condemning certain firm practices (price discrimination, consignment selling, retail price maintenance, exclusive dealing) which *can* be (they need not be) efficient means of building or maintaining firm size.

While the law on collusion (e.g. price fixing) has not foundered on the shoals of market structure, this part of the *Act* has been interpreted within the intricacies of criminal conspiracy but without an overall articulate goal of the *Act*. Some cases have, therefore, at least to an

---

[13] See Footnote 5.
[14] The 1986 *Act* does not appear to have been 'captured' by 'business' in the sense that a number of cases since 1986 (mergers and others) do not demonstrate a desire by officials to maximize producers surplus. A number of business spokesmen suggest that the bureaucracy (and courts) are not interpreting the 'intent' of the 1986 Act. For businessmen (or others) to rely on some implicit intent is clearly unreasonable. In addition, the Act does have real teeth.

economist, collapsed in a pile of legalist absurdity.[15] Several amendments have been made to this section of the *Act* in the last fifteen years to narrow the licence of judges.

The Economic Council of Canada in its 1969 analysis of Canadian competition law argued for a coherent base of economic efficiency (static and dynamic) for the law. The 1986 *Act* does underline efficiency as one goal of competition policy. Three years of experience with the new Act is too short a time frame to generalize on the direction of the law. However, there is still likely to be a dichotomy between criminal provisions (conspiracy and a set of provisions such as bid rigging and price discrimination) dealing with price setting and civil proceedings dealt with by a new specialized tribunal (The Competition Tribunal). The 1986 *Act* drastically rewrites merger and monopoly law. The former is dealt with by a case-by-case behaviourial analysis to determine whether the merger lessens competition substantially. However, jurisdiction over mergers is divided between what was originally an investigatory authority (the Director of Investigation and Research) and the Competition Tribunal. Tensions are evident and damage the application of the law. Monopoly provisions have been replaced with an 'abuse of dominant position' analysis which while relying somewhat on Article 86 of the Treaty of Rome concentrates too much on examining exclusionary practices against rivals.

In what follows, I provide a brief history of the development of anti-combines law in Canada and then examine the important sections of the law attempting to explain in economic terms why the law developed the way it did. An examination of the Canadian experience indicates the difficulties in:

(a) mandating what are anticompetitive business practices under a criminal code,
(b) proving economic definitions (such as the definition of the relevant market) in criminal cases,
(c) the difficulties in forming an effective coalition to amend the law, especially in a federal state,
(d) the tensions in formulating antitrust policy in small open economies.

---

[15] See Section 3. A number of authors do not feel that the conspiracy section has been interpreted too narrowly. For example see Webber (1982).

## 2. HISTORY OF THE LAW

### 2.1. Overview

Between 1889 and 1920 there were few cases and fewer convictions. With no specialized enforcement arm, there was little scope for the Department of Justice or normal criminal law enforcement service (the police) to ferret out anticompetitive business offenses. In the 1910 amendments, the law was altered to introduce a system of public complaints which is still in practice today. 'Any six or more citizens ... might make an application to a judge for an order directing an investigation.'[16, 17]

Public awareness was considered to be an important feature of early versions of the *Act*, a view that if people were aware of the offence, the violation would somehow disappear; a somewhat liberal and antiquated view which was accompanied by minor fines. Until 1975, the Act legislated the maximum fine for most offenses. Unlike the motto of Gilbert and Sullivan, the Canadian *Acts* for many years did not 'make the punishment fit the crime'.

The Canadian laws state that conspiratorial agreements (anyone who 'conspires, combines, *agrees* or arranges') are illegal if they lead to an '*undue* lessening of competition'. Until the new 1986 *Act*, mergers and monopolies had been illegal if they acted to the '*detriment of the public*'.

In essence, 'undue lessening of competition' is a less onerous burden for the prosecution to prove since the courts have assumed that a lessening of competition in merger and monopoly cases can occur without a corresponding public detriment. This occurred because the courts spent little effort in defining competition. Competition was a loosely-felt 'natural right' which would result in competitive prices,

---

[16] Kaiser 1987, p. CAN 1-19; see also Bureau of Competition Policy 1989 p. 7.

[17] The judge had to hold a hearing within 30 days and at that hearing applicants and alleged violators appeared and provided evidence. If the judge concluded that a *primae facie* case had been made as to an offence, he would order an investigation. The investigation was carried out by a three-person board—one person appointed by the applicant, one by the alleged violator and a third by these two board members. Two of the Board had to sign a Report which was duly published in the *Canada Gazette*. This public announcement of the offence was largely the penalty. A continuation of the offence was a crime, however no additional enforcement or surveillance mechanism was provided.

adequate choice and quality. The courts have not had the issues or thoughts of allocative or technical efficiency clearly in mind. I can find no case where economic efficiency is mentioned never mind calculated. Because the courts did not have a strong normative economic vision, they relied on their own legalistic interests. As a result, the words 'agrees' and 'undue' were sometimes interpreted in ways that economists would consider odd as was 'detriment to the public'. For example, several cases decided in the late 1970s and 1980s suggested that a price-fixing agreement was only 'undue' if it virtually eliminated all competition. A large (near 100%) section of the market had to be involved, acts of communication beyond simple oligopolistic acceptance of tacit collusion had to exist and the practice had to be shown to have anticompetitive, detrimental effects.[18]

In the 1980s important changes were made to the *Act*. For conspiracy, i.e. price fixing, words were added to the effect that competition did not have to be extinguished for an offence to exist. The law on mergers and monopolies was the subject of much more major changes —pre-notification for mergers; a means of obtaining advance rulings of the legality of mergers; a new quasi-judicial Competition Tribunal to examine civil remedies; and the substitution for the monopoly sections, the offence of the abuse of power by a dominant firm. These changes have been valuable but not without problems.

## 2.2. Administration of the Law

The 1923 *Act* introduced a permanent Registrar. Either he or a specially appointed Commissioner would carry out investigations.[19]

In 1952, two separate functions were created. The Director of Investigation and Research, analogous to the Registrar in the past, had the authority to initiate any inquiry (as well as respond to the six

---

[18] See *R. v. Atlantic Sugar Refineries Co.* [1980], 2 S.C.R. 644, 115 D.L.R. (3d) 21, 54 C.C.C. (2d) 373, rev'g 91 D.L.R. (3d) 618, 41 C.C.C. (2d) 209, [1978] Que C.A. 25, rev'g 26 C.P.R. (2d) 14 (S.C. Crim. D.V.).

*R. v. Aetna Insurance Co.* [1978], 1 S.C.R. 731, 75 D.L.R. (3d) 332, 34 C.C.C. (2d) 157, rev'g 62 D.L.R. (3d) 447, 22 C.C.C. (2d) 513, 12 N.S.R. (2d) 362 (S.C. App. Div.), rev'g 52 D.L.R. (3d) 30, 19 C.C.C. (2d) 449, 12 N.S.R. (2d) 416 (T.D.).

[19] In 1935, a permanent three-person Commission was established; this part of the Act was held to be *ultra vires* in 1936 by the British Privy Council (due to the ability of the Commissioners to approve rationalization agreements).

citizens' complaints). A separate Commissioner, the Restrictive Trade Practices Commissioner (RTPC) would hear the Director's case and could recommend prosecution in a public Report to the Minister.[20] Prosecution was before normal courts. In 1976, the RTPC, then a three-person Commission, was empowered to itself determine in a hearing the legality of certain practices—tied selling, refusal to sell and price discrimination—the first step in creating a quasi-judicial tribunal separate from the court system.

The 1986 *Act* attempts to more clearly demarcate the investigatory and prosecution functions from the adjudication function; provides new powers to the Director in negotiating changes to proposed mergers (see Section 5); replaces the RTPC with the Competition Tribunal and establishes that Tribunal as the specialized court to enforce the *Competition Act*. The Competition Tribunal is composed of both judges and lay-professionals.

The penalties for illegal activities include public awareness, fines and jail terms; prohibition orders are common. Two sections of the Act have never been utilized—the lowering of tariffs and the break-up of the firm (although the former had been recommended in a number of RTPC reports).[21]

Until 1952, the maximum penalty for a conspiracy offence was a paltry 10,000 and several judges stated that the fines were merely nuisances, 'a very trivial license to commit crime'.[22] In 1952 the maximum penalty was raised, but not enormously. In 1975, the ceiling was removed on the penalty for most offenses but left at $1,000,000 for conspiracies. Only in the past few years have substantial fines been viewed as important deterrents as the 1986 *Act* raises the ceiling in conspiracy cases to $10 million.

---

[20] One Director resigned (McGregor) in 1948 because a Report (Flour Milling) was not tabled by the Government for a year. As a result of this a Royal Commission was appointed to study the Combines Law, and substantive recommendations were suggested including the requirement that all Reports by the Director must be tabled in the House of Parliament within 30 days.

[21] The examination of mergers under the 1986 *Act* has included the impact of lowered tariffs under the Canada-USA Free Trade Agreement.

[22] McRuer in *R. v Northern Electric, Co. Ltd. et al.* (1956), 116 C.C.C. 98 at 99 (Ont. S.C.).

## 3. COLLUSIVE ACTIVITIES

Section 45 (Conspiracy) of the 1986 *Competition Act* is close to the section as it was written in 1920–1923, is still criminal law and is thus enforced by the criminal law system. As only 'undue' agreements are illegal, the offence is not under a *pre se* rule. Until the mid-1970s, this section of the *Act* was consistently interpreted, in the main, by finding an undue agreement if a significant share of the industry was involved. The courts shifted their interpretation in the late 1970s and early 1980s to find competition limited unduly only if competition was virtually stifled. Legislators responded by altering the wording of the relevant sections of the *Act* to weaken the required proof.[23]

### 3.1. Proof of an Agreement

The finding of whether an agreement exists rests on whether facts, direct or circumstantial, indicate that the parties intended to agree.[24] Until the late 1970s the Courts held that it was not necessary that the parties intended to limit competition unduly, 'intention is only relevant to the question of agreeing.'[25]

In most cases the court does not have direct evidence (for example a written document outlining the conspiracy) and must infer an agreement from circumstantial evidence.[26]

Dunlop *et al.* 1987 argue that the Canadian courts have attempted (unsuccessfully) to distinguish between 'the unwritten but explicit agreement', and the 'tacit or unspoken agreement'. An unwritten but explicit agreement is inferred by the Courts when the parallel acts or communications between the alleged conspirators are far greater than one would 'normally' assume[27].

---

[23] As of 1976, bid-rigging is *per se* illegal. (Section 47 of the 1986 *Act*.) There is no test as to whether bid-rigging (a conspiracy to set prices in a process of bidding) is undue, the existence of the conspiracy is sufficient for condemnation.

[24] See Gosse 1962, p. 101, 102.

[25] Ibid., p. 102.

[26] See *Container Materials Ltd. et al. v. R.* [1942], S.C.R. 147, [1942] 1 D.L.R. 529, 77 C.C.C. 129.

[27] In the mid-1970s, two alleged conspiracy cases were dismissed at a preliminary hearing. Cement producers and aluminum producers were charged by the Crown with

A 1976 case, *R. V. Canadian General Electric Co. Ltd. et al.*[28] demonstrates the kinds of evidence which suggests to the courts that an agreement existed. In this case, three large firms (CGE, Sylvania, Westinghouse) had 95% of the Canadian large lamps (industrial) market. The three firms sold to dealers who resold to the retail market, as well as selling directly to final buyers mainly under sealed-bid tenders. The evidence showed that prior to 1959 there was 'extensive price cutting in the industry'.[29] The evidence that convicted the three included the three firms putting new identical pricing plans into effect simultaneously in 1959 which fixed the discounts according to customer classifications and size of order. In 1961 Sylvania and Westinghouse adopted CGG's revised pricing plan announced at the annual meeting of the Canadian Electrical Distributors Association. Other evidence suggested a 'conspiracy' to the Courts. First, a sealed-bid tender to the City of Montreal produced 27 identical bids from dealers who purchased from the three producers; second, the companies monitored *retail prices*; and third, agents were instructed to withdraw tenders made below the list price.[30] Thus the Courts inferred that the *substance* of the communications between firms,[31] the complexity of the pricing plans which were identical, and the results (lack of price cutting which was prevalent before the alleged agreement) were proof that an agreement existed[32].

---

price fixing. The accepted defense in both cases was price leadership without collusion, or conscious parallelism. *R. v. Canadian Cement Lefarge Ltd.* (1973), 12 C.P.R. (2d) 12 (Ont. Prov. Ct.). *R. v. Aluminium Company of Canada Ltd.* (1975), 22 C.P.R. (2d) 216. In the former case, Judge Camblin in dismissing the case at a preliminary enquiry stated. 'Defence counsel, I am sure, will agree that the Crown has established that the companies involved are using a base freight factor pricing system which did not come about by mere coincidence . . . On reviewing the evidence the Court is of the opinion that the resulting prices set by the companies are the result of conscious parallelism and the companies are therefore discharged.'

[28] (1976), 15 O.R. (2d) 360, 75 D.L.R. (3d) 664, 34 C.C.C. (2d) 489 (HCS).

[29] *Ibid.* at p. 376 O.R., p. 680 D.L.R., pp. 504–5 C.C.C.

[30] In one instance such an instruction was marked 'note and destroy'. (D.L.R. at 690) (Obviously, it was not destroyed.) In another case, where C.G.E. received a contract by cutting the price below list, the C.G.E. sales manager wrote Westinghouse to apologize and stated that the gross profit would be given to charity! *Ibid.* at 693.

[31] Mr Justice Pennell stated 'Why should there be need to inform competition of an error made within one's own marketing system', and 'why is it necessary for a competitor to report the "misdemeanours" of a rival's agents to its own principals' (*Ibid* at 695).

[32] Another 1976 case assists us in understanding the meaning of an anticompetitive

Thus a charge of conspiracy was levied when the quality of evidence suggested an agreement and when the conspirators controlled a substantial part of the defined market.

The *Atlantic Sugar* case of 1980, added ambiguity to the Canadian Courts' finding of an 'agreement'. The three companies argued (successfully) that each *independently* adopting, *tacitly*, a policy of limiting price cutting to that required to 'maintain historical market share' was not an illegal agreement to limit competition.[33] Nearly identical market shares and a combined 98 to 100% share of the market were maintained for 25 years.

The Trial Judge showed a peculiar economics intuition in stating:

> 'Thus by natural osmosis the price of an homogeneous product tends to reach the same level *but the [competitive] process might be costly and is certainly inefficient.* There are two ways to avoid it. First, by the members of the industry conspiring to fix prices which is illegal, or by the members of the industry making a conscious effort to parallel the price of the leader.' (26 C.P.R. (2d) 14 at 96, emphasis added.)

and

> 'Since the Act does not prohibit a member of an industry from taking into account and following his competitors' price changes, be they up or down, it follows that he is not prohibited from taking into account and following *the system* upon which price changes are made. (*Ibid.* at 100, 3101).'

---

agreement. The Armco case involved ten corporations accused of price fixing in the metal culverts market; a market involving the sale by sealed-bid tender to municipalities. The Court stated that an agreement must involve 'the mutual arriving at an understanding or agreement.' The alleged conspirators had formed the Corrugated Metal Pipe Institute after a period of substantial price competition. The Institute adapted an 'open price' policy; all members were encouraged (but not required) to publish price lists, including discounts and to make these price lists available to all—customers, competitors or the public. One firm, Robert Steel, published such a list. No one followed Robert's prices. Six months later, the same firm published a new price list. All members of the Institute accepted this price list as the basis for submitting tenders. This acceptance proved, to the Court, the mutual arriving at an agreement, even without any incriminating or supporting evidence other than the acceptance of a common price list. See *R. v. Armco Canada Ltd.* (1975), 17 C.P.R. (2d) 211 at 259 (Ont. S.C.), affd. (1976), 24 C.P.R. (2d) 145 (C.A.). 13 O.R. (32) 183 at p. 42, 70 D.L.R. (3d) 287 at p. 296, 30 C.C.C. (2d) at p. 192.

[33] *Supra* at p. 657 S.C.R., p. 30 D.L.R., p. 382 C.C.C.

It is clear from these quotes that the judges were not considering consumers surplus or competition but the behaviour of competitors *vis-a-vis* each other.[34]

### 3.2. Definition of Undue

As the discussion of *Atlantic Sugar* indicates, the crux of the Canadian conspiracy law rests on the meaning of an *undue* lessening of competition.

Until 1977, a conspiracy limited competition unduly if it involved firms with control of the market.[35] Although the courts have not defined control explicitly they do examine market shares. Markets tend to be narrowly defined, e.g. English language daily newspapers in the Maritime provinces of Canada.[36] There was much discussion of the meaning of 'undue' in each case and as to whether it involved a 'single'-or 'double'-intent standard. The single-intent standard was that the conspirators did intend to agree; the double standard required both the intent to agree and the conspirator's knowledge that the agreement would have an undue effect.[37]

The problem is that the courts did not articulate basic economic premises, namely that if the conspirators undertook the costs of organization, then they themselves assumed they had market power.

---

[34] In the *Fertilizer* case [1981] 2 WWR 693 (Alta. S.C.)., the Crown alleged that seven producers of fertilizer in Western Canada had conspired to limit competition through a set of enabling factors which produced identical prices in an industry characterized by substantial cost difference and locational advantages. The enabling factors were the establishment of zone pricing (which is price discrimination when transport costs are significant as in this case), the standardization of terms of sale, joint refusals to sell to buying groups and the exchange of products between the producers. The Crown lost the case because the price lists it relied on as showing identical prices were not in fact transactions prices. The accused provided evidence that substantial price cutting below list price occurred and that real competition existed. Mr Justice Brennan was not impressed by arguments as to oligopolistic interdependent behaviour '. . . it is a standard and sound business practice for each to make or attempt to make itself aware and knowledgeable and as soon as possible of moves or decisions made by the others . . .' (at 161)

[35] Webber 1982 argues that the interpretation of the law did not change significantly in the late 1970s.

[36] *R. v. K. C. Irving Ltd.* (1976), [1978] I.S.C.R. 408, 72 D.L.R. (3d) 82.

[37] Under the Sherman Act, an agreement to limit competition is illegal. While there is no necessity of proving an undue lessening of competition, the proof required to find an agreement likely brings the Sherman Act close to the Canadian law.

While the courts accepted that one need not examine the impact of the conspiracy on prices or output, and that the 'good intentions' of the firms were not a defense, they did not accept that an undue conspiracy could occur without substantial market power.

This is best seen in a 1957 Supreme Court of Canada decision in a case involving fine paper mills and merchants[38] (emphasis added):

> 'In essence the decisions ... appear to me to hold that an agreement to prevent or lessen competition in commercial activities of the sort described in the section *becomes free to carry on those activities virtually unaffected by the influence of competition*, which influence Parliament is taken to regard as an indispensable protection of the public interest; that it is the abrogation to the members of the combination of the power to carry on their activities *without competition* which is rendered unlawful; that the question whether the power so obtained is in fact misused is treated as irrelevant; and that the Court, except I suppose on the question of sentence, is neither required nor permitted to inquire whether in the particular case the intended and actual results of the agreement have in fact benefited or harmed the public'.[39, 40]

A number of cases decided between 1957 and 1977 did not revolve around a 'monopolistic' tendency, market shares as low as 60% led to convictions yet the exact meaning of undue was unclear. In 1975, the *Combines Investigation Act* was revised to clarify Parliament's intent:

> '32(1). . . it shall not be necessary to prove that the conspiracy, combination, agreement or arrangement, if carried into effect, would or would be likely to eliminate, completely or virtually, competition in the market to which it relates or that it was the object of any or all of the parties thereto to eliminate, completely or virtually, competition in that market.'

Even with this 1975 amendment which seemed to articulate the

---

[38] *R. v. Howard Smith Paper Mills Ltd*. (1957), 8 D.L.R. (2nd) 449 at 473 also [1957] S.C.R. 403, [1957] 118 C.C.C. 321, affg [1955] O.R. 713, [1955] 4 D.L.R. 225, 112 C.C.C. 108 (C.A.), affg [1954] O.R. 543, [1954] 4 D.L.R. 161, 109 C.C.C. 65 (H.C.I.).
[39] *Ibid*. 4 D.L.R. 161 at 189.
[40] Mr Justice Spence also dismissed potential or poised competition as irrelevant. 'If the public had to rely on this distant possibility its protection would be slight indeed.' (*Ibid*. at 195).

'intent' of Parliament, two cases in the late 1970s again raised the issues present in the 1957 *Fine Papers* case, the *Atlantic Sugar* case, already discussed, and *Aetna Ins. Co. v. R.*[41] In these two cases, the Supreme Court of Canada appeared to return to a double-intent standard before finding an agreement undue.

The Quebec Supreme Court stated in *Atlantic Sugar*:

> '... none of the refiners was obliged to compete more strongly than it felt desirable in its own interest. Each refiner was entitled to decide not to seek to increase its market share as long as this was not done by collusion ... That this involved a lessening of competition is apparent. However it is equally clear that it did not involve a *suppression of competition*' (46 C.P.R. (2d) at 31, emphasis added.)

'Aetna *et al.*' were a set of fire insurance underwriters who composed the Nova Scotia Board of Insurance Underwriters, 55 to 70% of the number of firms operating in the market (defined as providing fire insurance in the province of Nova Scotia), and 70 to 80% of the number of policies over a ten-year period. The Board set prices. The trial court acquitted the firms of lessening competition unduly. The Nova Scotia Supreme Court reversed the judgement. The Supreme Court of Canada returned the acquittal in a significant 5 to 3 majority opinion. The majority held that there was still competition in the industry and that '... competition was not *stifled* by the boards actions ...' (emphasis added).[42-44]

The Canadian law on conspiracy was again amended in 1986 to deal first with what constitutes an agreement and second as to what constitutes undue.

Section 45(2.1) now reads as follows:

> '... the court may infer the existence of a conspiracy ... from circumstantial evidence, with or without direct evidence of communication ...'

---

[41] See footnote 18.

[42] 75 D.L.R. (3d) at 346.

[43] The minority held that the meaning of 'undue' had nothing to do with alleged benefits or whether elements of competition still remained. 'It is not an ingredient of the offence [conspiracy] that proof must be made that competition was in fact lessened unduly.' (*Ibid.* at 338)

[44] A 1983 decision by a lower court (not appealed) rejected the double-intent standard. *R. v. Thomsom Newspapers Ltd.* (28 October, 1983, unreported, Ont. H.C.J.)

This addition to the law does not make conscious parallelism an offence, it merely emphasizes that the finding of an *agreement* does not necessarily require direct evidence.[45] Another section was added:

> '45(2) ... it is not necessary to prove that the parties intended that the conspiracy, combination, agreement or arrangement, have an effect set out in subsection (1).'

While 45(2) appears to dispose of a double-intent standard, the magical word 'competition' has yet to be defined.

### 3.3. Exemptions

Over the years various market activities have been implicitly or explicitly exempt from the conspiracy provisions of the Act.[46] Until 1976 the sections of the Act were held to cover only goods or articles not services. It was not until 1976 that services were directly covered.[47] Explicitly excluded from the conspiracy section were labour relations, fisherman, shipping conferences, security underwriters, amateur sport, professions, regulated industries and export agreements. The list of exemptions requires some explanation. Easiest to explain are labour relations (collective bargaining) and underwriting (agreeing on an issue price of a security). Fishermen are likely a specific form of labour relations. Amateur sport ostensibly does not involve fees; export agreements are a common form of mercantilistic conspiracy against foreigners. Exemptions for professions and regulated industries are due to effective lobbying for the former as well as provisions of the Canadian Constitution. Under the *Constitution Act* provinces can regulate business activity (professions are also regulated provincially). It has been held that the *Combines Investigation Act* does not relate to an activity undertaken by a group acting in accordance with a valid provincial regulatory statute.[48, 49] This exemption has held not only for

---

[45] See Dunlop *et al.* 1987, p. 137.
[46] An excellent discussion is found in Kaiser and Nielsen-Jones, 1986.
[47] The explanation for the 87 years of neglect could be either indifference or a desire to exclude all financial services. Banks were explicitly exempt.
[48] *Jabour v. Law Society of British Columbia et al.* (1980), 44 C.P.R. (2d) 36 (B.C.C.A.); (1980) 45 C.P.R. (2d) 160 (B.C.S.C.).
[49] Crown agencies (corporations or agents of a government) were expressly exempt until 1976.

the activities of a regulated profession but also for an industry whose price is regulated by a government authority.[50]

## 4. MONOPOLY

### 4.1. Overview

The 1986 *Competition Act* does not mention monopolization as an offense. As in the European Community, the Canadian law makes 'Abuse of Dominant Position' an offense. This offence is a substantial modification of past Canadian antimonopoly law.

A brief history of the monopoly provision is in order. The basic prohibition against mergers and monopolies was placed in the law in the *Combines Investigation Act* of 1910, where an illegal monopoly must have '... operated or be likely to operate to the detriment or against the interests of the public, whether consumers, producers or others.'[51, 52]

The same definition of a monopoly offense was adopted in 1923 into the new *Combines Investigation Act*. Until the 1952 amendments establishing the Director of Investigation and Research and the Restrictive Trade Practices Commission, there was no formal mechanism to study monopolies, and one case to that date (1910–1952).[53]

The 1960 amendments to the *Combines Investigation Act* separated the merger and monopoly provisions of the 1923 *Act* from each other and from combinations in restraint of trade, but did not reword the definition of a monopoly.

Section 33(2) of that *Act* defined monopoly as follows:

> 'Monopoly' means a situation where one or more persons either substantially or completely control throughout Canada or any area thereof the class or species of business or are likely to operate it to the detriment or against the interest of the public, whether consumers, producers or others ...'

---

[50] *R. v. Canadian Breweries Ltd.* [1960], O.R. 601; 33 C.R.I. (H.C.J.); 126 C.C.C. 133.
[51] S.C. 1910 C. 9, S. 2 (c).
[52] Monopoly provisions were present in the Criminal Code prior to 1910.
[53] The 1935 Amendments to the Act distinguished the merger and monopoly section from the combines section. (1935) Statutes of Canada, c. 54.

The key question is clearly how one defines 'to the detriment of the public'. The limited number of cases (one successful prosecution) is not much grist for the economists' mill as to the objective function implicit in this Section 33 and in its interpretation.

I would argue that the judges have used the principle of *exclusion* as their guide—no notion of monopoly deadweight losses appears anywhere. Concepts of consumers' or producers' surplus are noticeably absent. What is present is a search for the power that monopolization leads to, a power unavailable by other (e.g. conspiracy) means. This monopoly power is not simply the power to set the monopoly price since a cartel could do that, but the power to *exclude* rivals or products or imports.

Between 1940 and 1985 there were 11 monopoly prosecutions in Canada and eight reports by the Director of the Combines Investigation Branch[54] into suspected monopoly cases (there is some overlap between the reports and the prosecutions).[55] Of the 11 prosecutions, the crown had only *one* prosecution and conviction (Eddy Match (1951/54))[56], one guilty plea (Electric Reduction Company of Canada (1970))[57] and three prohibition orders (where guilt was not admitted).[58-60]

---

[54] The Combines Investigation Branch came into being in 1952.

[55] Kaiser 1984. Kaiser and Nielsen-Jones 1986, discuss several other monopoly charges by the Crown. In *R. v. Hoffman-LaRoche* (20 May 1976, Ont. Prov. Ct. [unreported] ), the charge was dismissed at a preliminary hearing. In 1985, the Crown and the defendant agreed to an order of prohibition (*R. v. Pacific Northwest Bus Co.* (1985), 6 C.P.R. (3d) 265 (F.C.T.D.) ). Rosenbluth (1979) discusses the monopoly investigations in 1929 and 1937 which did not issue reports and which concluded that no monopoly existed (Eddy Match, (1929) and National Sewer Pipe Company (1937). Rosenbluth discusses a 1946 price discrimination case against Canadian Industries Ltd. (1952, 104 C.C.C. 39) which 'had a monopoly of small arms ammunition which voluntarily changed its practices' (p. 331). Rosenbluth also discusses a number of 'merger for monopoly' cases in newspapers, most of which were tried under the price discrimination clause and are not reported here.

[56] (1952), 104 C.C.C. 39

[57] 61 C.P.R. 235 (H.C.)

[58] *Wee Folk Diaper Service Inc. et al*, Exchequer Court of Canada, February 10, 1971 (unreported); *R. v. Anthes Imperial Ltd.*, Federal Court of Canada, February 22, 1973 (unreported); Canadian Safeway 12 C.P.R. (2d) 3.

[59] The seven prosecutions that were lost include: *Staples et al. (Dominion Fruit and Lumber Co.* (1940) (unreported); *B. C. Sugar* (1960) 38 C.P.R. 177; *Canadian Breweries* (1960) 126 C.C.C. 133, *Allied Chemical* (1976); *R. v. Hoffman-Laroche Ltd.*, Ont. Prov. Ct. May 20, 1976 (unreported); *Canadian General Electric, Canadian Westinghouse and Sylvania Electric* (1977) *(Large Lamps)* 34 C.C.C. (2d) 489 (H.C.); *K. C. Irving* (1977) 32 C.C.C. (2d) 1; *R. v. Thomson Newspapers Ltd. et al.* (28 October, 1983, unreported, Ont. H.C.J.).

[60] Of the eight reports issued by the Commission, one resulted in a prosecution and a

The *Eddy Match* case, a 1951 conviction, represents the abuse of dominant position that has become the basis of the antimonopoly law since 1986. *Eddy Match Co. Ltd.*, together with four other companies which it controlled, were convicted of monopolizing the business of wooden matches production to the detriment of the public through a set of 'abusive', predatory' or *exclusionary* practices supposedly leading to or flowing from its dominant position. None of these practices were criminal in nature, 'that these practices were employed merely adds to the evidence leading to a conclusion that the control exercised by appellants was likely to operate to the detriment or against the interest of the public'.[61] The abuses (or means used for exclusion) included predation[62]; the use of fighting brands (new brands underpricing an entrant's prices); market loading (increasing supplies (thus decreasing price)) wherever entry occurred; resale price maintenance and industrial spying.

In other monopoly cases (the Canadian Safeway prohibition order, and the two most 'famous' cases lost by the Crown—*Canadian Breweries*[63] and *B. C. Sugar*,[64] both mergers for alleged monopoly control) the Courts concentrated on the power that monopoly yielded in *addition* to that of a cartel. In the *Canadian Breweries* case, the Judge stated

> '... the evil, whatever it be the effect on competition or anything also which constitutes the offence as here charged, must be shown to flow from this merger and not from *collateral acts* which might have been the subject of another charge ...'(126 C.C.C. 133 at 8, emphasis added)

and

---

guilty plea. (*The Phosphates Case* (1966) *Electric Reduction Company of Canada* (1970). A second report resulted in a prohibition order, without a guilty plea (*The Cast Iron Soil Pipe Case* (1967), *Anthes Imperial Limited* (1973) ). A third report resulted in a change in firm behaviour (*The Ammunition Case* (1959), *Canadian Industries Limited*). A fourth report led to a prosecution and an acquittal (*The Large Lamps Case* (1977, *Canadian General Electric, et al.* (1977) ). Prosecutions were not attempted in the remaining three reports (*Sudbury/Copper Cliff Newspapers* (1964), *British Columbia Propane* (1965), *Zinc Oxides* (1958) ).

[61] *Eddy Match Co. et al. v. The Queen,* (1954) 109 C.C.C. (p. 20).
[62] No evidence of sales below marginal cost was discovered.
[63] *Supra* footnote 50.
[64] *R. v. British Columbia Sugar Refining Company Limited and B. C. Sugar Refinery Ltd.* (1960), 32 W.W.R. 577 (Man. Q.B.).

'... under the Combines Act it must be demonstrated beyond a reasonable doubt that the merging of competitive corporations is likely to put it within the power of the merger to so *extinguish competition as to affect prices* by *monopolistic control* ...'. (*ibid.* at 27, emphasis added)

In the *B. C. Sugar* case, the Judge found that evidence—as to the existence of price fixing, the adaption of the basing point system, pricing in Canada up to the world price plus tariff, the refiners' use of quotas to beet-sugar growers—were all in existence before the merger and therefore not an offence of the merger.[65] The Judge went on to say that the merger 'must also establish excessive or exorbitant profits or prices'[66] and that 'the Crown must also establish a virtual stifling of competition'. (38 C.P.R. 177 at 637)

In the *K. C. Irving* case, the market was defined as 'English daily newspapers printed and distributed in the Maritimes'.[67] The court agreed that the firm had a monopoly in this market but held that the monopoly did not operate to the public detriment since the firm did not attempt to alter the editorial policy of the newspapers nor were advertising rates higher than in other areas of the country[68].

This case exemplifies the distinction between conspiracy and monopoly cases (at least until the conspiracy decisions of the late 1970s). A conspiracy was unlawful if it had market power; the actual practices or behaviour of the conspiracy were not an issue. A monopoly however

---

[65] Mr Justice Williams did not accept the arguments that these enabling factors resulted in higher prices. 'I am asked to draw the inference that the use of the basing point system has operated, or is likely to operate, to the detriment of the wholesalers, jobbers or manufacturers, and therefore, indirectly, to the rest of the public because it resulted in higher prices. This is simply argument: There is no evidence from which I can draw any such inference.' 32 W.W.R. at 612.

[66] In a conspiracy case discussed earlier (*R. v. Howard Smith Paper Mills et al.*), Mr Justice Spence stated that a conspiracy could not be judged as undue or not on the basis of the level of prices or profits.

[67] *Supra*, footnote 36.

[68] The *Large Lamps* case was brought under both the conspiracy and monopoly sections of the Act—*R. v. Cdn. General Electric et al.* (1977), 34 C.C.C. (2d) 489 (H.C.), 15 D.R. (2d) 360. As noted earlier, the three firms were convicted of a conspiracy to adopt similar sales plans, maintenance of prices and segmenting of the market, including a scheme of consignment selling. The Court determined that the three firms had 95% of the market and that while '... they assert the necessary market control to constitute a shared monopoly ...', the firms were not guilty of operating a monopoly to the detriment of the public. The court held that the alleged detriments were the conspiracy and not the results of the shared monopoly.

was only 'undue', i.e. operated to the detriment of the public, if its *behaviour* was undue. Undue behaviour consisted of either explicit exclusionary acts or practices beyond those attributable to conspiracy.

### 4.2. The Monopoly Provisions Under Competition Act of 1986

The 1986 *Competition Act* includes a new practice, 'abuse of dominant position', as a matter reviewable by the Tribunal.[69] Abuse of dominant position occurs when the following three actions *all* occur:

> 71(1)(a) the firm(s) 'substantially or completely control' a class or species of business in Canada or any area thereof;
> 
> 79(1)(b) the firm(s) engage in a practice of anticompetitive acts, nine of which are explicitly listed in section 78 of the *Act*;
> 
> 79(1)(c) the practice of anticompetitive acts has had, is having or is likely to have the effect of preventing or substantially lessening competition in a market.[70]

The first of these is a structural test; the second, behavioural while the third examines the results of the combination of structure and behaviour.

A defence is allowed—superior competitive performance (79(4)). Dominant position is not illegal under rights flowing from the Copyright Act, Industrial Design Act, Patent Act, Trade Marks Act or '... any other Act of Parliament pertaining to intellectual or industrial property [which] is not an anticompetitive act' (79(5)) or if the practice occurred more than three years earlier and has ceased (79(6)). Besides prohibition, divestiture is possible as a remedy (79(2)).

This section of the *Act* is very different from the previous *Act*, very different from the US *Sherman Act*, Section 2, and based on Article 86 of the Treaty of Rome.[71]

The 1986 *Competition Act* adds explicit *behavioural* criteria to the existing structural interpretation of a monopoly—does the firm (or group of firms) *abuse* its dominant position so that there is a substantial lessening of competition. However the set of acts listed in Section 78 includes acts which are truly competitive and, relies on *exclusion* of

---

[69] For an excellent description and legal interpretation of this section of the Act, see MacDonald, 1987.

[70] Sec. 79(1)(c).

[71] See the companion paper by George and Jacquemin, 1990.

rivals as its principle. The *Act* also incorporates an *intent* test. These changes are thus potentially troublesome.

The problem for anticombines policy is to be able to distinguish competitive, non-collusive acts from abusive acts. This distinction is most difficult since the intent of competition is to 'dominate' rivals. It is rivalry which promotes efficiency and lowers prices. Unfortunately, the precise wording of proscribed behaviour in Sections 78 and 79 of the *Act* could result in efficiency-enhancing acts being proscribed. For example, consider clause 1 which renders illegal, when the three conditions (Section 79(1)(a), (b) and (c)) are fulfilled, 'squeezing a competitor's margin for *the purpose of* impeding the competitor'. Assume that firm A is vertically integrated to the retail level and that A also supplies B, a non-vertically-integrated retailer. A raises the wholesale price to B; A does not raise its retail price. Is this done for *the purpose* of driving B from the market? What evidence does a Court accept (even a specialized Court such as the Competition Tribunal) to prove this intent? If A can survive on a lower retail margin than B, is this because A is more efficient, because vertical integration is more efficient or because A is engaging in predation? Utilizing exclusion of rivals as a principle would appear to repeat the errors of the past.[72]

Part of this subsection, (and other parts of the *Act* as well) rely on the intent of the acts as a means of separating lawful desirable price cuts from illegal price decreases. Utilizing an intent standard can have serious deleterious economic repercussions unless the interpretation and application of the law recognizes subtle economic nuances which has not been the case under the Canadian law in the past. Arguably, this section of the law is a practice reviewable by the specialized Competition Tribunal and not the criminal courts and thus economics should have some influence on decisions.

## 5. MERGERS

### 5.1. Overview

As I have indicated in previous sections, mergers have been regarded relatively ambiguously in Canada. Horizontal mergers may increase

---

[72] The provisions regarding abuse of dominant position in the EEC's Treaty of Rome refer not to exclusionary practices but to unfair practices aimed at customers or suppliers. (See George and Jacquemin, 1990.)

market power but there is an underlying notion (and truth) that Canada is a small economy and a significant trader in a world market.[73] As a result, mergers, (as with specialization agreements since 1986)[74] are viewed as creating tensions or trade-offs between potential market power and potential increased efficiency.

Against this backdrop of the underlying notions of the role of mergers marches the jurisdictional drumbearer. Mergers in Canada, pre-1986, had to be declared anticompetitive under criminal law. As a result, the Crown had to prove that the merger would (or did) 'substantially lessen competition to the detriment of the public'; that the act of merger would (did) raise prices, for example, by conferring market power unavailable to the parties to the merger prior to the amalgamation (the *B. C. Sugar* case discussed above). Since economists have conflicting theories of why firms merge, differing views as to the gains from merger, and have difficulties in proving that a merger would (or did) unambiguously increase price over what would have been in the absence of the merger, it is not surprising that *no* convictions were achieved under the law in its pre-1986 form.

The 1986 Canadian antimerger law borrows heavily from the approach adopted in the USA. First, it relies on civil law. Secondly, a pre-notification requirement is featured as is a system by which the Director of the Bureau of Competition Policy can provide binding advance ruling certificates. Finally, however, unlike the US law an efficiency exemption is written into the law.

### 5.2. Merger Law Prior to 1986

In the 1910–1986 period, there were only 9 merger prosecutions, 7 resulted in acquittal, 2 were settled, not by a trial, but by a guilty plea.[75]

In *R. v. Canadian Breweries*,[76] the Crown following a Report by the RTPC in 1955, alleged that Canadian Breweries' acquisitions of 37

---

[73] Acquisition of a Canadian business by a non-Canadian requires an additional process contained in the *Investment Canada Act*, 1985 c. 20.

[74] Sections 57, 58, 62 of the *Competition Act* allow for the registration of Specialization Agreements which enable real efficiency gains (including a significant increase in the real value of exports or a significant fall in imports) that are greater than *and* offset any lessening of competition.

[75] Goldman, 1988.

[76] *Supra*, footnote 50.

breweries (chiefly in Ontario) reduced consumer choice and produced a monopolistic industry. The court stated:

> '... I do not think it is an offence against the Combines Act for one corporation to acquire the business of another merely because it wishes to extinguish a competitor. It is not the nature of the merger that is important, but what is important is whether it has operated to the detriment of against the interest of the public, or is likely to do so.'[77]

The court then stated that detriment to the public would occur if there was 'substantially' little competition left and that in this case there was no offense since '... it cannot be said that the merger has given it a monopoly in the beer market either in any one Province or in Canada as a whole.'[78] The firm's highest market share was 51.8% in Ontario.

In *B. C. Sugar*, (examined earlier) the Court held also in 1960 that a merger which led to a firm with 100% of the market (British Columbia and Manitoba beet-sugar refining) was not detrimental to the public because of potential competition and because the merger did not increase the market power already exercised by the firms. That market power already led to prices at the world price plus the Canadian tariff. How could a monopoly then increase prices? There were then no mergers which would be detrimental to the public. The decisions in *B. C. Sugar* and *Canadian Breweries* nailed the coffin shut on merger cases in Canada.[79, 80]

Prior to the mid-1970s an anticompetitive conspiracy was undue if the court was convinced that a substantial share of the market entered into an agreement to fix prices. Mergers and monopolies were however

---

[77] *Ibid.* at 23.

[78] *Ibid.* at 32, emphasis added. An additional point which was important in the decision was the fact that retail prices of beer were controlled by the provinces; the provinces also controlled the number of sales outlets.

[79] Kaiser and Nielsen-Jones, 1986, state that '... if one could prove complete control of a market as a result of a merger, one could *presume* detriment ... the accused were acquitted in the two cases because of a lack of substantial or complete control in the relevant markets.' (p. 425).

[80] In *R. v. Thomson Newspapers Ltd.* (1983 unreported), the Ontario Supreme Court acquitted two chains involved in a partial merger and simultaneous closing of newspapers in a number of cities on the basis that '... detriment must be found in addition to the finding of a lessening' (of competition). (Quoted in Kaiser and Nielsen-Jones, 1986, p. 426.)

anticompetitive only if they virtually excluded rivals, eliminated competition *and* operated to the detriment of the public.

Gosse argues that the phrase 'to the detriment of the public' in merger and monopoly cases 'should have been interpreted in the same way as they did 'unduly' [in conspiracy cases]' (Gosse, 1962, p. 184). It was not, nor should it necessarily have been.[81] If Gosse's opinion had held, mergers would have been declared illegal if they involved a substantial share of the market. A strict structuralist approach to merger policy in Canada's small open economy would have been wrong. However, the law which forced a burden of proof not possible to fulfil in an *ex ante* setting,[82] interpretations which simply considered increases in market power rather than the ability to continue existing practices and no clear concept of either the role of competition in the economy or the impact of mergers on competition[83] together produced an unworkable antimerger law.

### 5.3. Merger Law since 1986

The 1986 *Competition Act* follows the US. Hart-Scott-Radino Antitrust Improvements Act[84]. The *Act* also enshrines Oliver Williamson's classic 1968 article on 'Economies as an Antitrust Defence'.

In essence, the *Competition Act* gives divided jurisdiction over mergers to the Director of Investigation and Research of the Bureau of Competition Policy and to the Competition Tribunal. Only the Director can bring cases to the Tribunal and to this point in time seven have been taken. The Director can allow mergers to proceed without applying to the Tribunal. The latter process has been the course followed in 99% of the cases. Critics state that the law is now one of

---

[81] In the *Canadian Breweries* case, the court accepted the view that the two phrases were identical. The court argued in *K. C. Irving Ltd.* that this view was incorrect (*supra* footnote 35 at 10).

[82] 'The Criminal nature of the provisions appeared to have been at the root of the problem . . .' Kaiser and Nielsen-Jones, 1986, p. 423.

[83] Both the OECD, 1984, and Khemani, 1989b, suggest that Canadian merger law had adopted a behavioural rather than a structural test in determining whether mergers were anticompetitive. Strictly speaking that view is correct. However, judicial interpretation of the pre-1986 law was markedly influenced by structure since merger for monopoly appeared to be an initial step in evaluation. The post-1986 merger provisions do rely on behavioural criteria.

[84] 15 U.S.C.A. 1311; Pub. L. No. 94-435, 90 Stat. 1383 (1976).

administered justice with no public inquiry and no real knowledge of the Director's criteria.

Parties to a merger must pre-notify the Director of their intent

If   (a) the parties, and affiliates have total assets or annual revenues in excess of $400 million (Canadian).
and (b) the transaction has a value of over $35 million (Canadian).

Notification is also required if in addition to criteria, a) and b) above, a party acquires more than 20% of the voting shares of a public corporation or 35% of all the shares or assets of a public corporation or raises its share from 20% (or 35%) to over 50%.

A merger is now judged anticompetitive if it 'prevents or lessens or is likely to prevent or lessen, competition substantially.'[85] If the Director is convinced that the merger is not anticompetitive and that he has insufficient grounds on which to apply to the Tribunal for an Order (the criteria for judgement are discussed below), then he can provide an *Advance Ruling Certificate* (ARC), or an *Advisory Opinion* (AP). The Certificate precludes the Director from applying to the Tribunal for a remedial order if the merger takes place within one year of the Order and if circumstances are as attested to the Director in the pre-notification procedure.

The Director has various options if he is of the opinion that the merger will contravene the *Act*. First, he can allow the merger to proceed under a program of compliance, monitoring its effects. In some of these cases, the Director provides an AP. Importantly, the advisory opinion can be subject to certain undertakings (i.e. voluntary modifications to the merger). While the AP is *not* binding on the Director or his successors it does provide a degree of comfort to merging parties. Second, the Director can apply to the Tribunal for a Consent Order approving a negotiated settlement between the Director and the parties. Third, the Director can apply to the Tribunal for a remedial or Contested Order—i.e. involuntary modifications to the merger or dissolution.[86]

---

[85] In the remainder of this section, I concentrate on horizontal mergers. However, the *Act* covers vertical mergers (sources of supply s. 92(b), outlets s. 92(c) and other types of mergers s. 92(d) ).

[86] In a Consent Order hearing, the Director and the parties to a merger have an agreement and are applying for an enforceable order; the voluntary undertakings that parties provide to the Director in an AP are not 'clearly' enforceable.

### 5.4. Problems in the Merger Law

The intent of the law appears to some to have been that where an advance ruling certificate was not issued, and where the merger required modification, the Tribunal would examine the merger.[87] However, this option has been largely precluded for two reasons. First, firms usually cannot wait for a lengthy Tribunal examination to take place before knowing whether the merger will be allowed or not. Second, the Tribunal has rejected two of four Consent Order applications thus relatively increasing the value of an ARC, a negotiated settlement with the Director (Palm Dairies Ltd., Imperial Oil Limited).

The Tribunal recently has twice ordered modifications in the Imperial Oil Consent Order and in an oral ruling lambasted the Director's office for inadequate analysis and preparation. There is clearly a 'turf' war under way but since the Director is the only party who can apply to the Tribunal for a Consent Order in a merger case, the future role of the Tribunal may well be negligible.

### 5.5. Lessening Competition Substantially

At first glance, Section 64(1) of the 1986 *Competition Act* appears to repeat the law which proved unenforceable up to 1986. To an economist there does not appear to be a significant difference between the old phrase 'to the detriment of the public' and the new phrase 'prevent or lessen competition substantially' (SLC). The differences however are important.

First, merger law is now under the civil not criminal code; while the substance of the law does not change, the burden of proof does change appreciably. Second, and crucial, the arbitration of SLC is not the normal court system but the specialized Competition Tribunal (or, as we have seen, most often the Director himself). Third, the new phrase 'SLC' is closer to the undue lessening of competition test in conspiracy law than the old 'detriment to the public' test in merger and monopoly cases under the pre-1986 *Act*.

The *Competition Act* states that the Tribunal cannot find that a merger is anticompetitive 'solely on the basis of evidence of concentration or market share' (Section 91(2)). Thus, a pure structuralist

---

[87] See Grover and Quinn, 1989.

analysis of the impact of mergers on competition is precluded. However, any examination of SLC must begin with market definitions and market share. The Director's staff are clearly using a case-by-case analysis with little emphasis on rigid structural criteria. (See Khemani, 1989b.)

The *Act* lists a set of factors which may be considered by the Tribunal in determining the anticompetitive aspects of a merger. These include the extent of foreign and domestic competition (s. 93(a), (c), (e)) entry barriers (s. 98(d)) and whether the acquired firm is failing (s. 93(b)).

Section 96(1) encapsulates the 'Williamson efficiency exemption' into the law.

> '96. (1) The Tribunal shall not make an order under section 92 if it finds that the merger or proposed merger in respect of which the application is made has brought about or is likely to bring about gains in efficiency that will be greater than, and will offset, the effects of any prevention or lessening of competition that will result or is likely to result from the merger or proposed merger and that the gains in efficiency would not likely be attained if the order were made.'

The 'efficiency' defense has not, to this point, proven to be an important ingredient in decisions. The Bureau has stated that '... efficiency claims advanced by merging parties should be examined with extreme care if not conservatively (Khemani, 1989a, (p. 17)), ...'. 'In most cases efficiency claims have been revised downwards by Bureau and Competition Policy Staff' and '... there has as yet been no case where these efficiencies have had to be weighed in and of themselves against the substantial lessening of competition arising from the merger'.[88] The section sees gains in producers surplus as important. However, it does not make explicit the nature of the trade-off nor does it articulate how real efficiency gains are to be measured.

The only material made public as to the nature of the alterations in mergers negotiated between the Director and the parties to a merger are contained in press releases by the Director's office, speeches and the Director's Annual Report—for example, the Director undertook 146 examinations in 1987/88. The Director's Annual Report describes *10*

---

[88] Khemani, 1989a, p. 17, 18, 22.

of these examinations. Of the 10 described, 6 were transactions which involved no alterations to the terms of the mergers, 4 involved negotiated settlements, the only 4 involved restructuring of the deal. Those four were:

>  Canadian Safeway Ltd/Woodward Stores Limited.
> Nabisco Brands Canada Ltd/Interbake Foods Division of George Weston Limited
> Nestlé Enterprises Limited/Nabisco Brands Canada Ltd and General Foods Inc./Nabisco Brands Canada Ltd
> Trailmobile Group of Companies Ltd/Fruehauf Canada Inc.

Each is described in a paragraph. Lawyers, academics and businessmen have little to rely on in formulating opinions as to the likely consequences (from the Director's view) of prospective mergers. It is clear that changes are required in the operation of this section of the *Act*—either cases should be decided by the Tribunal, thus codifying law, or detailed published analyses should be provided by the Director[89].

## 6. VERTICAL RESTRAINTS

In the USA the application of the law on vertical restraints has changed significantly under the influence of recent academic research which suggests that many so-called 'restraints' may be efficient operations of markets, not anticompetitive behaviour. (See Mathewson and Winter, 1984; Dunlop *et al.*, 1987; McKee and Bassett, 1976)

Until 1975, the only vertical restraint covered in the Canadian law was resale price maintenance (RPM). RPM itself was added to the law in 1951 as a per se offence, Canada being the first country to prohibit resale price maintenance. This provision has remained essentially unchanged in the 1986 *Act* (the offence is now called price maintenance), and makes it illegal to attempt to influence prices upward or to discourage price reductions by the supplier or others. Both horizontal and vertical RPM are illegal. A suggested minimum retail selling price

---

[89] The new Director, Howard Wetston, has stated that detailed criteria will be published. In addition, the Bureau has increased significantly its provision of details on merger settlements.

is *prima facie* evidence of such influence unless the minimum is only a suggestion. Refusal to supply as part of a means to influence prices upward is directly illegal (61(2)).

Between 1954 and 1986 there were 108 RPM cases which led to prohibitions and/or fines, two-third of these post-1975. (Kaiser and Nielsen-Jones, 1986, p. 439.) The Courts have consistently upheld the RPM section, not enquiring into 'efficiency rationales' or whether the desire for RPM came from manufacturers or retailers. A policy of contributing towards advertising costs as long as the advertised price was not below dealer's cost was held illegal, (*R. v. A&M Records of Canada Ltd* (1980), 51 C.P.R. (2d) 225 (Ont. Co. Ct.)) as was a policy of advertising rebates tied to prices at suggested retail prices (*R. v. Church & Co. (Canada) Ltd* (1980), 52 C.P.R. (2d) 21 (Ont. Prov. Ct.)) *I. R. v. Epson (Canada) Ltd* (unreported, Dec. 11, 1987. Ont. Int. Ct.). The economics expert for Epson attempted to persuade the Court of the modern theory of vertical restraints, namely that RPM was necessary to maintain professional dealer integrity and service and to prevent free-riding. The Court said that the *Act* did not allow a defense, that influencing price was illegal *per se*.

The only defense allowed in Canada is that the practice is not an attempt to influence the price. The Courts have held that a manufacturer simply informing buyers of a suggested retail price is not an offense (*R. v. Philips Electronics Ltd* (1981, 2 S.C.R. 264, 59 C.P.R. (2d) 2124, 126 D.L.R. (3d) 767).

As noted earlier, in 1976 the RTPC was given jurisdiction over a set of 'reviewable practices' including exclusive dealing, tied selling, consignment selling, and refusal to supply. I will review the law and cases on exclusive dealing and tied selling, leaving the other practices to other references.[90]

Neither exclusive dealing nor tied selling are illegal *per se*; they are only illegal if they meet three conditions—they are so significant in the market (engaged in by a major supplier or are widespread (s. 77(2))) that they limit or impede entry and expansion and as a result lessen competition.

There has been one case on exclusive dealing and two on tied selling. In 1977, Bombardier, the leading Canadian manufacturer of

---

[90] See Andy and Vanneen, 1989; Kaiser and Nielsen-Jones, 1986.

snowmobiles, was accused by the Director of illegal exclusive dealing since dealers who carried Bombardier were not permitted to carry other brands.[91] The case was dismissed by the RTPC for the following reasons: other snowmobile manufacturers were easily able to acquire dealers therefore there was no substantial lessening of competition at the retail level; there was also no substantial lessening of competition at the wholesale level, since Bombardier produced only 10% of the snowmobiles produced in North America.

The RTPC found that the Bureau of Broadcast Measurement (BBM) did illegally tie the provision of both radio and television audience ratings in a package preventing the expansion of another firm (A. C. Nielson Co.) to the detriment of competition.[92] BBM's economic witness's arguments did not convince the RTPC that tied selling would not be injurious.[93] Since BBM was a non-profit cooperative owned by the broadcast users of radio and television ratings, there could be no gain to them from tying. Tying can raise profits by allowing price discrimination. However, a cooperative would not price discriminate against itself in order to increase profits. In addition, the market involved aspects of a public good and a natural monopoly (i.e. the cost for one set of ratings is lower than for two or more sets; a user's 'consumption' of ratings does not lower the amount available to other consumers; transactions costs are lower if all in the market use the same ratings); the Court did not accept these arguments.[94]

This very brief survey suggests that vertical restraints codefied into criminal law (RPM) are illegal *per se*, while vertical restraints under the purview of the Competition Tribunal fall under a rule of reason approach with economic arguments of some potential influence.

**BIBLIOGRAPHY**

Addy, G. N. and Vanneen, W. L. (1988), *Competition Law Service*, Toronto: Canada Law Book Inc.

---

[91] *Director of Investigation and Research v. Bombardier Ltd* (1980) 53 C.P.R. (2d) 47 (R.T.P.C.).

[92] *Director of Investigation and Research v. BBM Bureau of Measurement* (1981), 60 C.P.R. (2d) (RTPC).

[93] The author was the witness.

[94] A second tied-selling case (*Director of Investigation and Research and Broadcast News Ltd* (25 October 1985) (RTPC) (unreported) involved the status of an intervenor and did not result in an order.

Bladen, V. W. and Stykolt, S. (1958), 'Combines Policy and the Public Interest: An Economist's Evaluation', *Frealmann's Anti-Trust Laws*.
Bureau of Competition Policy, (1989), *Competition Policy in Canada, The First Hundred Years*. Ottawa.
Director of Investigation and Research, (1988) Competition Act, *Annual Report, (1987)*
Dunlop, B., McQueen, D. and Trebilcock, M. (1987), *Canadian Competition Policy: A Legal and Economic Analysis*, Toronto: Canada Law Book Inc.
Eastman, H. and S. Stykolt (1967), *The Tariff and Competition Canada*, Toronto: MacMillan of Canada.
George, K. and Jacquemin, A. (1990), 'Competition Policy in the European Community'. This volume. Chur: Harwood.
Goldman, C. S. (1988), 'Mergers, Efficiency and the Competition Act', Text of a speech delivered to the Commercial and Consumer Law Workshop, Faculty of Law, McGill University, October 15.
Gorecki, P. K. and Stanbury, W. T. (1984), *The Objectives of Canadian Competition Policy 1888-1983*. Montreal: Institute for Research on Public Policy.
Gosse, R. (1962), *The Law of Competition in Canada*, Toronto: Carswell.
Grover, W. and Quinn, J. (1989), 'Recent Developments in "Merger Law",' The National Conference on the Centenary and Competition Law and Policy in Canada, Toronto, October 24, 25.
Kaiser, G. E. (1987), *World Law of Competition Unit A, Volume 3. Canada*, New York: Matthew Bender.
Kaiser, G. E. and Nielsen-Jones, I. (1986), 'Recent Developments in Canadian Law: Competition Law'. *Ottawa Law Review*, **18.2**, 401-517.
Khemani, R. S. (1989a), 'Merger Policy and Small Open Economics: The Case of Canada' in Dankbaar B. Gronewegen J. and Schenk H. (eds), *Perspectives in Industrial Economics*, The Netherlands: H. Kluwer Academic Publishers.
Khemani, R. S. (1989b), 'Merger Policy in Small vs. Large Economics'. Prepared for the National Conference on the Centenary of Competition Law and Policy in Canada, Toronto, October 25.
Krattenmaker, T. and Salop, S. (1986), 'Anti-Trust Analysis of Anti-Competitive Exclusion: Raising Rivals' Costs to Achieve Power Over Price', 90 *Yale Law Journal* 207.
MacDonald, B. (1987), 'Abuse of Dominent Position', *Cdn. Competition Policy Record*, **8(1)**, 59-75.
Mathewson, G. F. and Winter, R. A. (1984), 'An Economic Theory of Vertical Restraints, 15 *Rand J*. 27.
McKee, J. S. and Bassett, L. R. (1976), 'Vertical Integration Revisited', 19 *Journal of Law and Economics* 17.
OECD (1984), *Mergers and Recent Trends in Mergers*, Paris.
Prichard, J. R. S., Stanbury, W. Y. and Wilson, T. A. (eds), (1979), *Canadian Competition Policy: Essays in Law and Economics*, Toronto: Butterworths.
Skeoch, L. A. and McDonald, B. C. (1976), *Dynamic Change and Accountability in a Canadian Market Economy*, Ottawa: Supply and Services Canada.
Waverman, L. (1986), 'Abuse of Dominant Position. The Monopoly Provisions of the 1986 Canadian Competition Act', mimeo.
Webber, F. (1982), 'Oligopoly and the Combines Investigation Act', *Cdn. Business Law Journal*, **6**, 453-492.
Williamson, O. E. (1968), 'Economies as an Anti-trust Defence', *AER*, March, pp. 18-36.

# UK Competition Policy: Issues and Institutions

KEN GEORGE*
*University College of Swansea*

## 1. INTRODUCTION

Laws against monopolies have been on the statute book since the seventeenth century. In principle it has also been possible to control monopolistic practices by resort to the laws of conspiracy and restraint of trade. For various reasons, however, these laws were ineffective. At the end of the nineteenth century cartel arrangements were widespread and they were further facilitated and strengthened by government wartime controls.[1]

During the years immediately following World War I there is ample evidence of official concern at the possible abuse of market power by monopolies and cartels but any chance there might have been of converting this concern into an effective competition policy was shattered by the economic troubles of the inter-war period. Faced with declining markets and increased international competition, rationalisation of industry was the name of the game, and although the dangers of monopolisation continued to be recognised there was a widespread belief that competition was less effective than combination and cooperation in achieving the necessary restructuring of industry. This attitude is found in several official reports including the 1929 Balfour Committee Report on Industry and Trade. In considering the problem of eliminating inefficient capacity the Report commented . . . 'it seems unquestionable that this operation can often be performed more

---

* I am grateful to William Comanor, Alexis Jacquemin and Hans Liesner for helpful comments on an earlier draft of this chapter. The usual disclaimer applies.

[1] Swann, D. *et. al* (1974) *Competition in British Industry*, London: George Allen & Unwin.

speedily and "rationally" and with less suffering through the mechanism of consolidation or agreement than by the unaided play of competition.'[2] Official encouragement of cartels together with the various measures of tariff protection that were introduced in the 1920s and 30s allowed trade associations to tighten their grip on the economy. It was also an era of big combinations which included the formation of industrial giants such as ICI, and Distillers Co. Ltd.

A turning-point came with the publication of the 1944 White Paper on Employment Policy in which the government committed itself to maintaining full employment and also emphasised the importance of increased productivity in solving the country's economic difficulties. The fear was that these objectives might be jeopardised by the price-fixing and other restrictive practices of monopolies and cartels.

In 1948 the Monopolies and Restrictive Practices (Inquiry and Control) Act was passed, which set up the Monopolies and Restrictive Practices Commission with powers to investigate monopolies and cartels and also to report on whether these situations might be expected to operate against the public interest. In defining the public interest the Act instructed the Commission to take into account, among other things: the need to achieve efficient production and distribution, a more efficient organisation of industry and trade, and the development of technical improvements. But there was no suggestion that the best way of achieving these goals would be through maintaining competitive markets. Indeed there was no mention of the word 'competition' in the definition of the public interest.

The work of the Commission turned out to be that of inquiry rather than control. From 1948 to 1955 twelve references were made mainly concerning the restrictive practices of trade associations. In addition a general reference was made on *Collective Discrimination*. The report on the latter published in 1955 concluded that various discriminatory practices such as aggregated rebates, and collective agreements to withhold supplies, restricted competition and were generally against the public interest. New legislation was recommended but the Commission was divided on the approach that any new legislation should adopt. The majority recommended a general prohibition of restrictive

---

[2] *Final Report of the Committee on Industry and Trade*, Cmnd. 3282, London, 1929, p. 179.

agreements with specific provision for exemption, while the minority supported compulsory registration for case-by-case examination. It was the minority view that prevailed.

The 1956 Restrictive Trade practices Act took restrictive practices policy away from the Commission (which was renamed the Monopolies Commission), and established separate machinery to deal with restrictive practices—the Registrar of Restrictive Practices, and the Restrictive Practices Court. The Act established compulsory registration of restrictive agreements, and such agreements were initially presumed to be against the public interest. The legality of individual agreements was to be established by the Court, with the Act laying down the conditions under which an agreement could be upheld. The boldest action taken in the 1956 Act was the prohibition of collective resale price maintenance.

Competition policy was further strengthened with the passing of three further Acts in the 1960s. The Resale Prices Act 1964 introduced the general prohibition of individual resale price maintenance (RPM), and laid down a policy of compulsory registration with the possibility of exemption by application to the Restrictive Practices Court. The Monopolies and Mergers Act 1965 brought mergers within the scope of Monopolies Commission investigations, while the Restrictive Trade Practices Act 1968 brought information agreements within the scope of restrictive trade practices legislation.

Yet the 1960s also saw the re-emergence of the old doubts concerning the efficacy of competition in achieving efficiency goals. The election of a Labour government in 1964 ushered in a period of active government intervention in industry. The year before they brought mergers within the scope of competition policy the Labour government had set up the Industrial Reorganisation Corporation (IRC) whose job it was to encourage mergers—presumably those in the national interest, which would not have come about, or not come about quickly enough, without some official prodding and public money. Behind this initiative was the view that British companies were frequently too small to compete effectively against overseas competitors together with the conviction that market forces could not be relied upon to rectify such structural weaknesses.

The same ambivalence was evident in the government's attitude towards restrictive practices. In 1962 the National Economic Development Council (NEDC) had been established as an instrument to promote the growth of the economy. Its role was to examine the performance and plans of industry, to consider obstacles to faster growth and to seek agreement on ways to improve competitiveness and efficiency. One of its first actions was to produce a report on the economy in the course of which a number of industries were consulted to discover their plans for investment and output. Following this it was agreed that it would be desirable to establish consultative machinery on a more permanent basis with the result that several industry Economic Development Committees (EDCs) were formed. The formation of the EDCs was a reflection of a growing dissatisfaction with competition as a means of promoting structural change. The emphasis on co-operation and planning is illustrated in the following statement made by the Director General of NEDC relating to the perceived need in 1969 to divert more of the country's resources into investment and exports: 'It is not enough to switch resources into investment and exports. We have to shift resources into the right investments and the right exports. All twenty-one of [the EDCs] are working at this moment to identify where this shift of resources can be achieved, what are the right investments and which are the profitable exports.'[3]

A general complaint made by the EDCs was that many efforts to improve efficiency were being hindered by the provisions of the restrictive trade practices legislation, i.e. beneficial co-operative arrangements such as joint marketing efforts overseas and agreements relating to capacity reductions in declining industries were being put in jeopardy by the provisions of the 1956 Act. As a consequence the 1968 Restrictive Trade Practices Act allowed for the exemption of industry-wide rationalisation schemes and agreements for promoting industrial efficiency, though these provisions have hardly ever been used.

Looking at the broad sweep of competition policy in Britain it is clear that the achievement of greater economic efficiency has been a central issue. In pursuit of this goal it is accepted that monopolies may have beneficial results, that the vast majority of mergers will not be harmful, and that co-operation will sometimes be better than competition. With the exception of resale price maintenance, where the position since 1964 has been one of near *per se* illegality the official attitude towards competition has been equivocal. This was certainly so

---

[3] Catherwood, H F R (May 1969) 'The Planning Dialogue', in *National Westminster Bank Review*.

during the inter-war years and again during the 1960s and early 1970s. As the 1970s wore on, however, the advocates of planning were thrown into retreat and with the election of the first Thatcher administration in 1979 the emphasis was placed more firmly on competition. Even so in some respects, as for instance in the handling of the privatisation of public sector monopolies, the rhetoric extolling the virtues of competition, has been stronger than the policies.

Currently competition policy falls under four main Acts, and is enforced by four main competition authorities.

The *Fair Trading Act, 1973* replaced those of 1948 and 1965. The most important institutional change was the creation of a new office —that of Director General of Fair Trading. The Director General was empowered to make monopoly references subject to the approval of the Secretary of State. The functions of the Registrar of Restrictive Trading Agreements were transferred to the Director General so that the surveillance of monopolies, mergers and restrictive practices was brought together for the first time. The Monopolies Commission was reconstituted and renamed the Monopolies and Mergers Commission (MMC). The Act changed the definition of the public interest, including for the first time explicit mention of the desirability of competition. The *Restrictive Trade Practices Act, 1976* consolidated previous enactments relating to restrictive trade practices. The *Resale Prices Act, 1976* consolidated Part II of the 1956 Act which prohibited collective RPM and the 1964 Act which introduced the general prohibition, subject to exemptions, of individual RPM. The *Competition Act, 1980*, which was a new piece of legislation, enables the Director General of Fair Trading to investigate the anticompetitive practices of single firms and to make 'competition references' to the MMC. Previously the anticompetitive behaviour of single firms could only be examined as part of a fullscale monopoly reference. The 1980 Act also empowers the Secretary of State to refer public sector bodies to the MMC.

The four main competition authorities are the Office of Fair Trading (OFT), the MMC, the Restrictive Trade Practices Court, and the Secretary of State for Trade and Industry. The OFT monitors restrictive trade practices, monopolies and mergers. Its Director General (DG) takes restrictive trade practices to the Court unless otherwise directed by the Secretary of State. The DG also makes monopoly references and competition references to the MMC subject to the approval of the

Secretary of State. Only the latter has the power to make merger references and public sector references under Section II of the 1980 Act. The MMC investigates the monopolies, mergers, anticompetitive practices and public sector bodies that are referred to it. If it decides that a merger, for instance, may be expected not to operate against the public interest there is no power to stop it; similarly if there is no adverse public interest finding with respect to the conduct of a monopolist. In these instances the MMC effectively has the power of decision. In the event however of an adverse public interest finding the MMC can only make recommendations, the power to decide on remedies resting with the Secretary of State. The latter has considerable powers at his disposal including, for instance, the power to stop a merger or to prevent monopolies from pursuing a particular course of conduct. But he is not bound by the MMC's recommendations. He decides at his discretion whether to accept the conclusions and recommendations of the Commission.

The following sections discuss the main developments and policy issues as they apply to the four major policy areas—horizontal collusive behaviour, monopoly and the abuse of market power, mergers, and vertical relationships.

## 2. HORIZONTAL COLLUSIVE BEHAVIOUR

### 2.1. The Legislation

Horizontal collusive behaviour is dealt with under the Restrictive Trade Practices Act, 1976. The Act provides for registration with the Director General of Fair Trading of agreements between two or more persons in the production or supply of goods in which the parties concerned accept restrictions in respect of such matters as: the prices to be charged or recommended; the terms or conditions on which goods are to be supplied or acquired; the process of manufacture to be applied; the persons to or from whom the goods are to be supplied or acquired; and the areas or places to or from which goods are to be supplied or acquired.

The definition of an agreement is widely drawn, encompassing agreements which are expressed or implied as well as those that are written. Similar provisions are made in respect of designated services, and the Act also provides for the registration of information agreements, again both for goods and for services.

The Director General is obliged to refer every registered agreement to the Restrictive Practices Court unless he is discharged from doing so by the Secretary of State for Trade and Industry.

Registrability of agreements depends on whether the agreements contain restrictions of a *form* specified in the legislation, and not on whether the agreement has the *effect* of reducing competition. UK legislation is thus described as 'form-based' rather than 'effects-based'. Registrability also depends on whether an agreement falls under one of the exempted categories. The list of exemptions is a long one. It includes a wide range of professional, transport and financial services, and agreements relating to such matters as adherence to standards, trade-marks and patents, and wages, employment and conditions of work.

Registered agreements are presumed to be against the public interest and the burden of proof is on the parties to an agreement to demonstrate benefit. When referred to the Court the parties to an agreement have to satisfy the Court that each restriction has beneficial effects falling into one of eight categories (the so-called 'gateways'). In addition they have to satisfy the Court that these benefits outweigh any detriment that may exist from operating the restrictions. This requirement is known as the 'tailpiece'. There are eight gateways, seven of them dating from 1956; the eighth introduced in 1968. They are:

(a) where the restriction is necessary to protect the public against injury;
(b) where removal would deny the public other benefits or advantages arising from the restriction;
(c) where the restriction is necessary to counteract measures taken by a non-party to prevent or restrict competition;
(d) where the restriction is necessary to negotiate fair terms with a non-party who controls a preponderant part of the trade in such goods;
(e) where removal of the restriction would have a serious and persistent adverse effect on unemployment in an area in which a substantial part of the trade in question is situated;
(f) where the removal of the restriction would cause a reduction in exports;
(g) where the restriction is required for the maintenance of any other

restriction found by the Court not to be contrary to the public interest;
(h) where the restriction does not directly or indirectly restrict or discourage competition to any material degree in any relevant trade or industry, and is not likely to do so.

There has been much criticism levelled at these gateways. For instance, gateway (b), the most frequently pleaded, has been criticised for being too widely drawn, and it is difficult to see any good reason for including gateways (e) and (f) in the evaluation of restrictive practices. But though the gateways may appear to be excessively wide the number of successfully defended cases is remarkably small and this is due in no small measure to the general presumption and the sting in the tailpiece.

One of the most celebrated cases upheld by the Court was that of the Cement judgement in 1961 when the main price restriction and supporting restrictions were found not to be contrary to the public interest. The Cement Makers Federation (CMF) was established in 1918 and most of the restrictions which were tested in 1961 dated from the inter-war period. The most important restriction was that by which members agreed to charge scheduled delivered prices for cement. In addition to this basic restriction there were supplementary restrictions such as deferred rebates and rebates for bulk supply and ex-works collection. The main argument used by the CMF in defending the restrictions was that they conferred substantial benefit by keeping prices lower than they would otherwise have been by avoiding wasteful use of transport and by the planned expansion of capacity to meet demand. The Court was also convinced that the price agreement would continue to be operated with responsibility and restraint. The common prices and marketing arrangements were in operation until February 1987. At the time they were abandoned the OFT was investigating the case with a view to applying to the Court to have the judgement re-examined. The OFT's renewed interest in the cartel may well have been a factor explaining its demise. Another was the growing threat from imports especially from other European Community countries.

What of the appropriateness of a judicial process for dealing with restrictive agreements? One advantage is that it is open; the arguments on both sides can be heard by all those concerned and assessed by the Court (consisting of a judge and two lay members). However, a judicial system operates most satisfactorily where agreements are made illegal *per se*. This is not so for British policy which requires a great deal of weighing up of complex economic arguments in order to decide whether an agreement should be upheld or not. The British system is particularly dependent on economic interpretation because the gateways cover so many economic objectives. But even if the gateways were more narrowly focussed on matters concerning efficiency in production and distribution and the promotion of research and development the problem would remain, particularly that of determining what is best for *long-term* competitiveness. It is, for example, reasonable to expect that a restrictive agreement will generally have an adverse effect on the prices paid by consumers in the short run. The analysis of the long-run effects of an agreement, are however, much more complicated and less amenable to the judicial procedure.

Another problem is the time taken to reach a decision. It may be two years or longer from the date when an agreement is registered before the Court reaches a decision. This delay is clearly too long for potentially beneficial agreements relating, for instance, to co-operation in R&D where a quick decision may be essential if the project is to go ahead. The delay also means that there is no way of catching agreements that are intended to last for two years or so, for such agreements can be registered, operated for the period up to the date of the Court hearing and then abandoned.

For reasons such as these plus the slim chance of success, firms have been reluctant to defend agreements before the Court. Much of the work of the OFT therefore has consisted of negotiating with firms with the aim of removing any offending restrictions so that the agreement becomes sufficiently innocuous for the Secretary of State to discharge the Director General from his duty of referring the agreement to the Court.

### 2.2. Effectiveness of the policy

The Cement case referred to earlier is one of only a handful of cases that have been upheld by the Court. Over the period 1956–87, 886 goods agreements, plus a further 57 services agreements, were referred to the Court. Once some early judgements, such as the Yarn Spinners case, had been made against the trade only a small number of cases have been defended, and only eleven have been successful. These figures seem to indicate an impressive success rate in eliminating restrictive practices, but they tell only a part of the story. There are in fact problems with the policy in terms of coverage, detection and lack of penalties.

As to coverage, all restrictive agreements, apart from those that are exempt, are registrable, even when they have no adverse effects on competition. Indeed it is possible that the legislation may deter agreements that have beneficial effects. This may apply for instance to co-operative R&D or joint overseas marketing schemes where the resources required are large in relation to firm size or where the risks are unusually high. Again, by increasing the information available to decision-makers co-operative arrangements may facilitate a quicker and smoother adjustment of capacity to demand changes than would otherwise be possible. It is arguments such as these that have been advanced in particular by the National Economic Development Office.[4] A problem of course is that co-operative schemes relating to R&D, marketing, or capacity adjustments may also facilitate tacit or explicit collusion. Nevertheless it has to be conceded both on theoretical grounds and on grounds of experience that beneficial schemes of this kind can exist.[5]

While policy may deter beneficial agreements it also fails to capture many agreements with substantial anticompetitive affects. Avoidance of the law can occur because as long as an agreement is carefully worded it may be immune whatever its effects. Evasion of the law may also occur when the parties to an agreement with substantial anticompetitive effects realise not only that they have no prospect of successfully defending the restrictions but also that there is no penalty in the event of the unlawful agreement being uncovered. That this is a serious problem is illustrated by the fact that in August 1988 the OFT had to take measures to break up a series of price-fixing cartels involving several glass manufacturers and distributors. The OFT had been alerted by a customer who had claimed he was 'tired of being ripped off' (reported in the *Financial Times*, August 9, 1988). This is one of several restrictive agreements to be uncovered in the

---

[4] See Hughes, A. (1978) 'Competition Policy and Economic Performance in the United Kingdom', in NEDO, *Competition Policy*, London, 1978.

[5] One of the early exponents of the limitations of competition and the virtues of cooperation was G. B. Richardson: See *Information and Investment*, London: Oxford University Press, 1960; and 'The Pricing of Heavy Electrical Equipment: Competition or Agreement?' in *Bulletin of the Oxford Institute of Statistics*, vol. 28, 1966.

construction trades, other recent cases involving price-fixing and tendering arrangements among suppliers of ready-mixed concrete, road-surfacing materials, and reinforced steel bars.

These weaknesses have led to calls for reform and in particular the adoption of an effects-based system. Basically this would mean that agreements which had the purpose or the effect of restricting competition would be prohibited unless they were exempted. Agreements would be notified to an admistrative body which would also decide on exemption. The process of registration and evaluation would thus be vested in one body instead of two as at present. In order to *deter* anticompetitive conduct, especially tacit agreements which are difficult to detect, it would be necessary to strengthen the investigatory powers of the OFT and most importantly introduce penalties for those found guilty of serious anticompetitive practices. For the same reason, i.e. difficulty of detection, there is good reason to provide for private actions along American lines.

There would be several advantages in adopting an effects-based policy. The objective of legislation would be clear and explicit. The legal form of an agreement which has been a troublesome part of the existing legislation would be irrelevant, since the only relevant matter would be its effect on competition. It would not be possible therefore for anticompetitive behaviour to escape simply on the basis of the wording of an agreement. On the other hand, agreements which did not have anticompetitive effects would be freed from control. One of the major problems with the existing approach would thus be overcome.

The authors of the 1979 Green Paper 'A Review of Restrictive Trade Practices Policy' considered these arguments and rejected them.[6] They argued that the present system had worked reasonably well, that a change to an effects-based system would involve a prolonged period of uncertainty for industry which could prove damaging and that it would be difficult for the competition authorities to get evidence of the anticompetitive purpose of an agreement unless it was equipped with stronger powers than are now available to the Director General.

---

[6] *A Review of Restrictive Trade Practices Policy: A Consultative Document*, Cmnd 7512, HMSO, London, 1979.

## 2.3. The 1988 Green Paper

However, all the weaknesses of existing legislation were at last recognised in the 1988 Green Paper which announced that the 'Government have concluded that the coverage of UK law should be defined in terms of the effects of agreements and concerted practices on competition rather than in terms of their legal form . . .' (para. 1.5). The case for change concludes:

> 'In sum, our present system is inflexible and slow, too often concerned with cases which are obviously harmless and not directed sufficiently at anticompetitive agreements. The scope for avoidance and evasion considerably weakens any deterrent effect the system has and enforcement powers are inadequate. The requirement to furnish insignificant agreements is not only wasteful of official resources but imposes an excessive burden on firms.'[7]

The only puzzle is that these points which were just as valid at the time of the first major review of policy in 1979 were not recognised then. The only explanation offered is that UK businesses have gained experience with the operation of European Community (EC) competition rules, so that a change to the new system will be less disruptive and give rise to less uncertainty than would have been the case in 1979.

The review considered two main options for an effects-based law: a statutory prohibition of agreements with anticompetitive effects, or a system which bans agreements with anticompetitive effects after they have been found upon investigation to be against the public interest. The second option was rejected as being too weak. Two options for implementing a prohibition system were then considered: a general prohibition with exemptions or a series of specific prohibitions. The latter was rejected on a number of grounds:-the difficulty of defining the prohibited agreement; the fact that agreements falling just outside the legal definition of a prohibited agreement would escape even though they might be seriously damaging to competition; that even if a list of specific prohibitions could be defined it would still be necessary to allow for exemptions; and the disadvantage of not being in line with the EC policy. Compatability with EC law clearly weighed heavily in the review:

---

[7] *Review of Restrictive Trade Practices Policy*, Cm. 331, HMSO, London, 1988.

'The effects-based prohibition which the Government proposes has the added benefit of alignment with existing EC law for the sake of consistency and simplicity. Increasingly, UK companies must have close regard to EC competition rules. Much greater compatibility between EC and UK law than the present system affords will make the latter more easily comprehensible and workable for the business community.' (para. 3.15).

The proposal therefore is for a general prohibition of all agreements and concerted practices which have the 'object or effect of preventing, or restricting, or distorting competition'. It will apply to vertical as well as horizontal agreements and to tacit understandings as well as overt arrangements.

The law will be aimed in particular at 'hard-core' practices such as price fixing, collusive tendering, market sharing, agreements to prohibit or restrict advertising and other promotional activities, and various types of collective action including refusal to supply, and discrimination between customers. These practices are thought almost always to have severe anticompetitive effects and they are listed to emphasise the general thrust of the new policy.

The new legislation will incorporate a general exemption test which will allow all applications for individual exemptions (including those involving 'hard-core' practices) to be judged against the same principles. For this purpose Article 85(3) of the Treaty of Rome is taken as the obvious precedent. (See the section by George and Jacquemin, p. 206ff.). The burden of proof will be on the applicant.

The new legislation may also incorporate block exemptions in which case the Government would again as far as possible make UK policy compatible with that of the EC. The categories of agreement which are listed as possible candidates are:

(i) exclusive dealing and purchasing agreements, as long as there is effective horizontal competition between manufacturers and retailers and no barriers to entering either the manufacturing or retailing market;
(ii) patent licensing, intellectual property and know-how agreements which encourage the speed of technology transfer as long as they are not accompanied by wider restrictions, such as an obligation to buy quite different products, as a price for a licence;
(iii) R&D agreements where co-operation in R&D is essential and

where the potential for any reduction in competition is limited;
(iv) franchising agreements, again subject to there being little prospect of anticompetitive effects.

It is the Government's intention to remove those exemptions which have long applied to many sectors of the economy, especially the services sector.

The implementation of the new policy will be in the hands of a new administrative authority, which will combine the investigating and decision-making functions along the lines of the European Commission and the West German Federal Cartel Office. The new authority will decide whether an agreement is anticompetitive and caught by the prohibition; if it is, whether it qualifies for an exemption, and if not, what penalties to impose. Firms will be able to appeal to the Restrictive Practices Court against decisions on exemptions, and against penalties.

The new authority will have substantial powers, including powers to require the production of information and documentation and powers of entry and search—these proposed powers being dependent upon reasonable suspicion of a breach of the prohibition. The authority will also be able to impose fines up to a maximum of 10 percent of total turnover.

Finally, and as an added deterrent to cartel agreements, the Government proposes to provide for private actions.

### 2.4. Concluding Comment

Restrictive practice policy in the 1950s and 1960s had some success in sweeping away the most blatant forms of restriction and in creating a more competitive environment. However, this success was patchy. There was both evasion and avoidance of the law, many areas of economic activity were exempt, and above all, policy did not deter anticompetitive behaviour.

In the 1988 review the Government at last recognised these deficiencies. The only surprise is that it has taken so long to do so. There may well be some administrative problems as the new policy is introduced and it will take some time for case law to be built up. The new authority will have wide discretion in administering exemptions and it remains to be seen how sympathetic it will be to restrictions which are alleged to have beneficial effects, or indeed how often parties will seek exemption.

However, the new policy will have some important advantages over the old. It focusses on the anticompetitive effects of agreements; by introducing heavy penalties and the opportunity for private actions there will be more powerful incentives for firms to comply; and it brings UK policy closer into line with that of the European Community. But it would be a mistake to imagine that the new approach will deal completely even with rather blatant forms of collusion. As US and EC experience shows, even far-reaching search powers and penalties will not stop all collusive behaviour.

## 3. MONOPOLY AND THE ABUSE OF MARKET POWER

### 3.1. The Legislation

Monopoly legislation is governed by the Fair Trading Act 1973 and the Competition Act 1980. Monopoly and competition investigations are undertaken by the MMC following a reference which is normally made by the Director General of Fair Trading. The Secretary of State, however, has the power to veto a reference, though he has not yet exercised this power.

The Director General has a duty under Section 2 of the 1973 Act of collecting information about market structure and business conduct and of generally keeping under review commercial activities in the UK. Information gathered in the course of these duties, together with complaints received from firms and members of the public, form the basis on which a decision is made whether to refer a monopolist for investigation.

A statutory monopoly exists when one firm has at least 25 percent of the market of a good or service. Monopoly policy also extends to cover oligopoly via the 'complex monopoly' provisions of the 1973 Act which says that a market may be liable to investigation when two or more companies together have a market share of at least 25 percent and act in a way that prevents, restricts, or distorts competition. The investigation of an oligopoly depends on finding that a complex monopoly exists.

Usually the Commission will be asked to report on whether the monopoly situation operates or may be expected to operate against the public interest. The latter is defined in very broad terms. Although Clause 84 of the 1973 Act states that the Commission 'shall have regard to the desirability of maintaining and promoting effective competition' it also instructs the Commission to 'take into account all matters which appear to them in the particular circumstances to be relevant . . .'.

There is no presumption in the legislation that monopoly is bad and the onus is on the Commission to show that a monopoly operates against the public interest.

The Commission carries out a monopoly investigation within a time limit specified in the reference. This period was commonly 18 months to two years with the provision for an extension, but more recently the Commission would in most cases expect to complete a reference within a year. The inquiry is wide-ranging and covers the various dimensions of industry structure and the conduct and performance of the monopolist.

If the Commission reports that a monopoly exists and that facts found by them operate or may be expected to operate against the public interest they must specify the particular effects which they have condemned and may make recommendations to the Secretary of State as to possible remedies. The recommendations are not always accepted. For instance, in *Household Detergents* (1966) recommendations of substantial cuts in price and in promotional expenditure were not implemented, and in *Film Processing* (1966) a recommendation that import duties on colour film be abolished fell on deaf ears.

One of the criticisms of competition policy contained in the 1979 Green Paper on restrictive practices was that various anticompetitive practices which were not registrable under the restrictive trade practices legislation could only be dealt with as part of a full monopoly investigation. The authors of the Green Paper called for a more flexible approach for dealing with these anticompetitive practices. This was provided by the Competition Act, 1980. The Act provides for the control of anticompetitive practices in both private and public sectors and also for the investigation of the efficiency of public sector bodies and the possible abuse of a monopoly position by such bodies.

### 3.2. Definition of a Monopoly

Proof of the legal existence of a monopoly is straightforward, a monopoly being defined in law in purely structural terms as a firm with

25 percent of the market that has been referred. However, the reference market is not necessarily the economically relevant market, a point which defendants have often been at pains to point out, and a market share of 25 percent is only an initial screening service and doesn't automatically qualify a firm for investigation—some evidence of the possible abuse of market power is also needed.

The main problem in determining the existence of monopoly power is that of defining the relevant market in both product and geographical terms. In order to exert market power a firm must have a significant share of the market but this share can be varied enormously by adopting a broader or narrower definition of the product market or of its geographical extent. In addition it is impossible to define a particular market share that will allow a firm to exert market power since this will vary with other features of the market.

### 3.3. The Commission's Assessment of Monopolies

Most of the Commission's inquiries have concerned highly concentrated markets and indeed several of them have approximated simple monopolies. However, as mentioned earlier, there is no presumption that a high market share in itself is bad, and assessment is based on a wide range of conduct and performance variables, including prices and profits, advertising expenditure and anticompetitive practices.

Between 1956 and January 1989 the Commission produced 53 monopoly reports in the goods sector. In about 14 of them profits and/or prices were judged to be excessive. In the *British Oxygen* case (1956) it was the near complete monopoly and the limited financial risk facing the company that led to the conclusion that profitability was unjustifiably high. The risk factor was also important in the report on *Colour Film* (1966). The Commission drew attention to the fact that Eastman Kodak 'has suffered no serious setback throughout its history' and recommended that prices should be cut. The profitability of the companies in the *Librium and Valium* (1973) and *Contraceptive Sheaths* (1975) references was so high that a conclusion of excessive profits could hardly be avoided however efficient the companies particularly in view of their near total monopoly positions. However, the Commission has cleared several cases where the profitability of a dominant firm may appear to have been excessive. In *Cigarette Filter Rods* (1969) efficiency, high risk, and ease of entry were important factors justifying a high return on capital. Efficiency was also an important factor in explaining why the Commission was satisfied with the high profits in the supply of *Cat and Dog Foods* (1977).

As to advertising, the classic case is that of *Household Detergents* (1966). The industry was found to be highly concentrated with Unilever and Proctor & Gamble accounting for 90 percent of sales. Profit rates were much higher than the average for manufacturing—about twice as high for Unilever and nearly four-times as high for Proctor & Gamble. The Commission found that the aspect of the firms' behaviour that operated against the public interest was the extremely high level of advertising expenditure. Price competition had been largely replaced by competition in advertising at a level that deterred new entrants, protected the existing market structure and permitted the earning of excess profits. Although a number of markets have been investigated in which advertising is important, *Household Detergents* remains the only case in which the level of advertising has been found to be against the public interest.

Several monopoly investigations have revealed the existence of various forms of anticompetitive practices. Many of these have involved vertical relationships which are considered in Section 5. The following are some examples of anticompetitive practices which the Commission has objected to. In *Petrol* (1965) the Commission objected to the requirement that petrol retailers should sell the supplier's lubricants exclusively. In *Metal Containers* (1970) the Metal Box Company rented closing equipment on very advantageous terms provided that only cans supplied by the company might be closed on the equipment, a condition which restricted the customer's freedom to purchase from other suppliers. The practice which was condemned in the *Indirect Reprographic Equipment* (1976) case was the tying of toner to the supply of copying machines. In the investigation into *Animal Waste* (1985) it was found that the dominant company had engaged in predatory pricing aimed at a particular competitor.

### 3.4. Problems of Assessment

There are substantial problems involved in assessing the evidence collected in the course of a monopolies inquiry. Apart from the sheer volume of material to be handled there are particular difficulties which frequently occur. Some of them are technical in nature, others are to do with the problems of a changing market environment.

Large size is commonly defended on grounds of scale economies but the measurement of these economies is a technical matter demanding a great deal of information. The Commission has been accused of failing to deal with this aspect of monopoly in a systematic fashion. The problem, however, is not that the Commission is unaware of the importance of quantifying economies of scale but that it is never easy and sometimes impossible to get the necessary information. In part this is due to the presumption in the legislation against intervention. It is for the Commission to find that there are effects which operate or may be expected to operate against the public interest. There is little incentive therefore for firms to produce the detailed information required.

Evaluation of profit performance involves some sort of reference to the concept of a 'normal' rate of profit, and the Commission has frequently used as a benchmark the average return for all manufacturing activity. Factors such as risk, managerial efficiency, entry conditions and market power are then examined to see if a much higher than normal return, when it occurs, can be justified. This is, on the whole, handled as well as can be expected. However, monopoly policy has to be conducted within a dynamic framework in which firms and industries have different growth rates, and in which profits are necessary to attract finance for expansion, either by means of retentions or from external sources. The dynamic interrelationships between profitability, investment and growth are much more difficult to handle. The growth performance of a monopolist is sometimes taken into account, at least indirectly, in assessing profit performance, but there is also a case for attempting to relate in a more systematic way the profitability of the firm on the one hand and its planned growth rates and investment on the other. There are difficulties or course in doing this. In particular the evidence which the Commission would have to rely on would be forecasts prepared by the firm under investigation and there would be a tendency for the forecasts of growth and investment to be inflated. Still the Commission is not unaccustomed to dealing with exaggerated claims so this should not be a reason for failing to take investment plans into account.

Another problem related to the process of change is the extent to which high profits in one activity should be allowed because they are helping to finance expansion in another activity. This transfer of funds is a common feature of business activity and indeed is both inevitable and desirable in a changing world. The issue has not been fully faced up to in monopoly policy because it is a product market which is referred for investigation and not the firm. However, it may in practice be less of a problem than might be supposed because the Commission has shown itself willing to accept high profits so long as they can be justified by a high level of efficiency.

The suggestion has in fact been made that firms rather than individual products should be referred for investigation.[8] For large diversified companies this would mean a complete change in the type of investigation undertaken with the focus of attention shifting away from market structure and performance to such matters as organisational efficiency, investment appraisal, resource allocation decision-making, etc. This is the approach adopted in public sector references. However, these are cases where, typically, a single product or service, or a limited range of products and services, are supplied and where the degree of monopoly is typically greater than that found in the private sector. It is doubtful whether the approach could be applied as effectively to diversified companies in the private sector or indeed whether it would be appropriate to do so.

As a final illustration of the problems of assessment take the difficulty of judging the future course of competition. In *Colour Film* (1966) the Commission commented that 'a monopoly position founded upon Kodak's reputation with the public is not likely to be suddenly eroded'. But between 1962 and 1982 Kodak's market share of film processing fell from 63 percent to 25 percent: true this didn't happen suddenly but it is nevertheless a striking example of loss of market power.

### 3.5. Remedies

The Secretary of State has wide powers exercisable by Order of Parliament to compel a firm to take whatever action is necessary to remedy any adverse public-interest finding, including the power to order a company to sell assets. Almost invariably, however, the remedies applied are ones which control aspects of a firm's behaviour or

---

[8] See *A Review of Monopolies and Mergers Policy: A Consultative Document*, Cmnd 7198, HMSO, London 1978.

performances and the usual way of achieving this is by voluntary undertakings negotiated by the Director General. Undertakings may remain in force for a considerable period of time before a company is released from them. For instance, Kodak was not released from an undertaking concerning profit levels until 1984, eighteen years after the *Colour Film* report, and undertakings given by Kellogs following the *Breakfast Cereals* report in 1973 are still in force.

In the most recent cases there has been a noticeable tendency to avoid administrative controls if at all possible and to rely instead on new entry or measures designed to loosen up the market. In Tampons (1986), for instance, though it was recognised that profits were higher than they would have been if the dominant firm had less market power it was concluded that in the absence of any entry barriers 'the currently rewarding profit levels should act as a magnet to attract new entry'. In *Postal Franking Machines* (1986) although the MMC estimated a rate of return of 72 percent over a six-year period on the supply of reference products it did not recommend measures to control prices or margins but proposed instead to reduce or remove factors restricting competition. In *White Salt* (1986), however, it was concluded that the absence of effective competition was unlikely to be rectified by the threat of imports or of new entry and so direct price controls were recommended as the only workable solution.

### 3.6. The Effectiveness of Policy

How effective has policy been in finding solutions to the monopoly problem? Some light on this question is shed in a study of eight markets which were investigated by the Commission.[9]

The main conclusion of the study is that the Commission's criticisms and recommendations and the resulting undertakings given to the Government had only a minor impact. Where the monopolist had lost market share this was attributed in the main to competitive forces operating independently of the effects of policy, and the same conclusion is reached in those cases where, although market dominance was maintained, competitive pressures had increased. However, action

---

[9] Shaw, R. W. and Simpson, P., (February 1985) 'The Monopolies Commission and the Process of Competition', *Fiscal Studies*.

taken as a result of monopoly investigations was found in a number of cases to have contributed to increased competition by reducing entry barriers and impediments to expansion by smaller firms.

Another interesting point made in the research is the view that monopoly investigations came too late. For instance in markets such as petrol, fertilizers, washing powders, and colour film, the leading firms had enjoyed dominant positions for a very long time before they were referred. There is no doubt that there are many cases where a monopoly situation should have been investigated much sooner, but on this issue of the timing of a referral there is also something of a dilemma.

The concern of monopoly policy is with entrenched positions of market power and not with firms whose market power and monopoly profit is short-lived. High profits are an important part of the competitive process—in financing expansion, attracting entrants and encouraging the search for substitute products. And so long as there are no important barriers to competition high profits will be competed away. The time taken for the competitive process to work will vary across industries, depending on such factors as differences in technology and the ease with which close substitutes can be developed. But it will also be affected by obstacles placed in the way of the competitive process by dominant firms. It is with these factors which prolong positions of dominance that monopoly policy is concerned, and the problem is to decide how soon a position of dominance should be challenged. The evidence suggests that the UK competition authorities have generally waited too long.

A factor which has limited the effectiveness of policy is that it has eschewed almost completely a structural solution. The possibility of enforced divestment has been considered in a number of cases. In their report on *The Supply of Beer* (1989), for instance, the MMC concluded that in view of the powerful market position of the national brewers, basic structural changes were needed if the adverse effects stemming from their ownership of public houses was to be remedied. One of the Commission's main recommendations was that brewers should be limited to owning no more than 2000 public houses, the rest to be divested. This recommendation was subsequently amended by the Secretary of State, with the effect that brewers have to release from a tie half the pubs they own in excess of 2000. Still, an important measure of divestment was achieved. In general the enforced division of companies is regarded as impracticable. However, firms do divest themselves of branches and subsidiaries, and this practice seems to be becoming more common. It is surely appropriate that they should be forced to do so when this is clearly in the interests of consumers and can be achieved without loss of efficiency. In this regard it is particularly disappointing that the government has not made greater use of the possibilities of competitive restructuring as part of its privatisation programme.

A particularly thorny problem in this area of competition policy is the control of oligopolies. As mentioned earlier, investigation of an oligopoly depends on finding that a complex monopoly exists. Evidence of a certain uniformity of behaviour must therefore be found before a reference can be made. When a reference is made the Commission's investigations are limited to those practices which form the basis of the complex monopoly and to those companies that are party to them. But the real problem in the absence of a structural approach is that of finding an effective remedy. How can a group of oligopolists be stopped from behaving as oligopolists?

Finally, the 1980 Competition Act, which was intended to provide a more flexible means of dealing with the anticompetitive practices of single firms, has produced very disappointing results. Up to the end of 1987 23 investigations had been initiated and 6 of these were referred to the Commission. The overall impression is that with one or two exceptions the cases dealt with have been rather trivial. Either British industry is largely free from these restrictions or they are somehow evading the attention of the authorities.

UK experience illustrates well the difficulty of finding an effective remedy for monopolisation. This would suggest the wisdom of having a policy to prevent the emergence of dominant firms in the first place. The most obvious arm of such a policy is the control of mergers and it is to this topic that we now turn.

## 4. MERGERS

### 4.1. Introduction

That merger activity is a major factor in explaining the long-term increase in the level of concentration in UK industry is confirmed in a number of studies. Although the most extreme estimates, which show mergers accounting for more than the whole of the increase, must clearly be heavily discounted the consensus of opinion is clear enough. Typically, mergers have contributed for upwards of 50 percent of the increase in concentration over the relevant study periods and company samples.[10]

An increase in concentration does not of course mean that monopoly power has increased correspondingly. In the UK the increases in concentration of the 1950s and 1960s occurred at a time when a whole range of restrictive trade practices were being abandoned. Many mergers involving small and medium-sized companies may strengthen the competition faced by dominant firms and so reduce market power. Other mergers may be defensive—a response to increased competition from overseas, or to the market power of suppliers or customers.

Nevertheless, there is concern about the level of merger activity in the UK and this concern naturally heightens during periods of intense merger activity. The period 1984–89 has witnessed a huge merger boom with expenditure in real terms in all three years, 1986–88, well in excess of the previous peak years of 1968 and 1972. And in contrast to previous merger booms which were short-lived the 1980s boom has already spanned five years. During this latest merger boom several very large companies, many of them household names, were acquired, including Distillers, Imperial Tobacco, Debenhams, British Home Stores and Rowntree.[11]

### 4.2. Merger policy

Merger policy in the UK dates from the Monopolies and Mergers Act 1965. The current merger control provisions are contained in the 1973

---

[10] See, for instance, Aaronovitch, S. and Sawyer, M., (March, 1975): 'Mergers, growth and concentration', *Oxford Economic Papers*; Hannah, L. and Kay, J. A., (1977); *Concentration in Modern Industry: Theory, Measurement and the UK Experience*; London: Macmillan; Curry B. and George K. D., (March, 1983) 'Industrial Concentration: A Survey', *Journal of Industrial Economics*, vol. XXXI.

[11] For further details see George, K. D. (1988) 'The Extent, Nature and Causes of Mergers', in George, K. D. (ed): *Macmillan's Mergers and Acquisitions Yearbook*, Macmillan, 1988.

Fair Trading Act and despite government reviews in 1978 and 1988 the basis for policy has remained largely unaltered since then.[12]

A reference to the MMC may be made where a merger creates or enhances a monopoly (currently defined as 25 percent of the relevant UK market) or where the value of the assets taken over exceeds a specified threshold (currently £30 million). There are separate provisions for newspaper mergers and what follows is concerned with the general body of mergers excluding newspapers.

If a merger is referred, the MMC is required to report on whether it operates or may be expected to operate against the public interest. The guidance which is given for interpreting the public interest is extremely wide, and is the same as that applied in monopoly investigations as laid down in Section 84(1) of the Fair Trading Act. It states that 'the Commission shall take into account all matters which appear to them in the particular circumstances to be relevant . . .'

The onus of proof is on the Commission to demonstrate that a merger may be expected to operate against the public interest and in doing so it must be able to demonstrate specific adverse effects.

If the Commission reports that a merger may be expected to operate against the public interest the Secretary of State has discretionary powers to prevent the merger or to compel dissolution if the merger has taken place. In practice this is normally done informally; the Director General of Fair Trading obtains undertakings from the firms concerned not to go ahead with the merger.

The review of mergers policy that took place in 1988 recommended no radical changes. The review notes the criticism that merger control procedures have been time-consuming and inflexible, welcomes the progress made in cutting the length of MMC investigations and declares that some of the proposed improvements in working methods and internal procedures will require minor legislative changes. In addition, two more substantial legislative changes were announced. The first is 'a formal, though non-mandatory, pre-notification procedure' designed to secure the rapid clearance of mergers where there are no grounds for a reference. Those who decide to pre-notify their proposed merger to the OFT will be required to complete a standard question-

---

[12] *A Review of Monopolies and Mergers Policy: A Consultative Document: Cmnd. 7198, H M S O London 1978;* Department of Trade and Industry, (1988), 'Mergers Policy' London: HMSO.

naire setting out basic information about the transaction and about the business involved. In simple cases, and provided the proposed merger has been publicly announced so that third parties have an opportunity to register objections, the merger will be given automatic clearance if nothing is heard from the OFT within a short period—possibly four weeks. In more complex cases it will take longer for the OFT to assess the position and in these cases firms will be informed that the right to automatic clearance has lapsed. Mergers which are not pre-notified will remain liable to reference to the MMC for a period of up to five years. The idea of a mandatory pre-notification procedure for mergers above a certain size was rejected because of the additional burden both for the authorities and for business; because firms in any case often choose to inform the OFT of merger proposals; and because 'it has been comparatively rare for completed mergers to raise issues which justify reference ...', although as the review admits in the same sentence, 'there have been a number of references of completed mergers in the recent past'! (para. 3.7). Voluntary pre-notification is not new to UK merger policy. However, the procedures proposed in the 1988 review give greater incentives for firms to pre-notify and will improve the information available to the OFT.

The second major legislative proposal announced in the 1988 review is for the introduction of a mechanism whereby parties to a merger can give legally binding undertakings to meet competition objections as an alternative to a MMC reference. Undertakings might involve divestment of some of the assets of the merging companies or the post-merger behaviour of the new group. If the DG is satisfied that the undertakings remove a possible adverse effect on competition, he can advise the Secretary of State against making a reference. If an undertaking to divest assets is not honoured, the Secretary of State will have powers without an MMC investigation to require divestment. This addition to mergers policy gives statutory backing to an apparently growing willingness of predators to divest themselves of assets in order to avoid a MMC reference. Thus, in their (unsuccessful) bid for the Imperial Group, United Biscuits announced their willingness to sell the Golden Wonder snack food division of Imperial in order to reduce the merged company's share of the snack foods market to below 25 percent. In announcing its successful takeover bid for Distillers, Guinness agreed to sell several brands of whisky in order to avoid creating a dominant position in the UK market. Indeed the then-Chief Executive of Guinness is reported to have said that the divestment was in the public interest because it would stimulate competition in the UK market! And Dixons, in its unsuccessful bid for Woolworths, announced that it would dispose of the Comet electrical chain in order to avoid a possible reference to the MMC.

### 4.3. Permissiveness of UK policy

A factor that colours the entire approach to merger policy is the presumption that most mergers are beneficial or, at least, neutral in their effects on competition and efficiency, and only a small number are likely to be against the public interest and thus need to be investigated. In 1969 the minister responsible for merger policy stated that 'in general, mergers are desirable if they lead to better management or genuine economies of scale without eliminating workable competition. In my view more often than not in Britain mergers will fulfil this condition'.[13] This continues to be the government's view. Thus in the 1988 White Paper, 'DTI—the Department for Enterprise', it is stated that the policy towards mergers 'enables the great majority of proposed mergers and acquisitions which do not pose a threat to competition to be decided by the market, without intervention from official agencies'. Thus, in mergers as in monopolies legislation, the presumption is against public intervention.

The permissive attitude towards mergers is reflected more clearly in the fact that the onus of proof is on the Commission to show that a merger would be against the public interest, and also in the number of mergers which have been referred to the Commission. The figures show that over the period 1977–87, 57 references (including 10 newspaper references) were made out of a total number of approximately 2493 mergers covered by legislation, or 2.3 percent of all referrable cases, and less than 1 percent of all industrial and commercial mergers. There are a number of possible explanations for this permissive attitude.

First, there has been an uncritical association of efficiency with size irrespective of *how* large size is attained. Merger proposals are invariably supported by claims of efficiency gains due to economies of large-scale production, or economies in distribution and marketing.

---

[13] See Board of Trade, (1969) *Mergers: A Guide to Board of Trade Practice*, London: HMSO, annex 5.

However, these advantages are more easily claimed than realised and research into the matter tends to show disappointing results.[14] Generally, there has been a reluctance to accept the argument that if economies of scale are present they can be realised through the process of internal expansion.

Second, there is the view that the takeover mechanism is an efficient selection mechanism, weeding out inefficiency and concentrating resources in the most efficient firms. At best, however, UK research in this area gives only highly qualified support to this view. Studies show a tendency for acquired firms to be smaller and less dynamic than other groups and to have lower profitability especially in the years immediately preceding a takeover, suggesting perhaps opportunistic behaviour on the part of the acquirers. Acquiring companies tend on average to be larger and faster-growing than other firms but profitability comparisons lead to less clear-cut conclusions. When acquired and acquiring companies are compared, size is the clearest discriminator with profitability comparisons again showing mixed results. Whatever general tendencies are found are usually rather weak with substantial overlap occurring in the performance of groups of firms.[15]

Third, there is the argument that, because the future is uncertain and industrial structure is constantly adapting itself to changing circumstances, the success rate in predicting the effects of mergers is likely to be low. Since most mergers are likely to be pro-competitive or neutral in their effects it is best to replace merger control with control of the undesirable conduct of monopolies when such conduct appears.[16] There are a number of difficulties with this view. The argument that mergers are pro-competitive is not convincing when internal growth is a feasible alternative. Again, the empirical work on the effects of mergers is not reassuring. Most investigators in the UK have concluded that post-merger profitability performance has been worse than pre-merger performance in a small majority of cases. Certainly there is

---

[14] See, for instance, Cowling, K. *et. al.* (1980) *Mergers and Economic Performance*, Cambridge University Press.

[15] See, for instance, Singh, A., (September 1975); 'Take-overs, economic natural selection and the theory of the firm', *Economic Journal*, Meeks, G., (1977); *Disappointing Marriage: A study of the Gains from Merger*, Cambridge University Press, and Kumar, K. S. (1984) *Growth, Acquisition and Investment*, Cambridge University Press.

[16] See Littlechild, S. C. (1978) *The Fallacy of the Mixed Economy*, Hobart Paper 80, Institute of Economic Affairs.

usually a finding that a large minority of mergers are 'successful' on the basis of these tests and also that the quantitative decline in post-merger profitability is typically small. Nevertheless, the results do not support an optimistic view of the effects of mergers.[17] Studies based on share-price movements appear to give more reassuring results. Much of the evidence relating to share prices in the short period immediately before and after a merger show substantial gains to the shareholders of acquired firms and gains also, albeit on a much more modest scale, to the shareholders of acquiring firms.[18] However, if these share-price movements are supposed to reflect the potential for longer-term gains in efficiency and profitability, it is odd that there is so little direct evidence of such gains in studies based on company accounts. In this context it is interesting to note that when share-price movements are examined over a longer period the picture becomes less clear-cut. Much of the evidence then suggest that the short-term gains are not sustained and that the share prices of acquiring firms decline relative to the values that would have been attained had the shares performed as well as those of non-merging companies.[19] Finally, the experience of UK monopoly policy is also not reassuring when it comes to correcting abuses of market power. As already seen in Section 3 the UK authorities have only once, for instance, resorted to the structural remedy of divestment of assets as a means of correcting monopoly abuse.

Wherever possible, and especially where structural remedies are regarded as impractical, it is better to prevent the emergence of market power than to attempt to modify business behaviour and performance once such power has emerged, and this suggests the need for a strong merger policy.

---

[17] See for instance, Meeks, G., *op. cit*; Kumar, M. S. *op. cit*; Hughes Alan, (1988), 'The Impact of Merger: A Survey of Empirical Evidence for the UK', in Fairburn J. A. and Kay J. A. (ed) *Mergers and Merger Policy*, Oxford University Press, 1989.

[18] Firth, M. 'The Profitability of takeovers and mergers' *Economic Journal*, June 1979; and 'Takeovers, shareholder returns and the theory of the firm', *Quarterly Journal of Economics*, 1980; Franks J. R. and Harris R. S., (1986). 'Shareholder wealth effects of corporate takeovers—the UK experience 1955-85' *London Business School Working Paper*.

[19] For a summary of the evidence see Hughes, A., *op. cit*.

## 4.4. Proposals for reform

Several criticisms have been made of UK merger policy. In the space available, attention is focussed on three of the most important ones—procedures for making references and applying remedies; the public-interest criterion and the onus of proof.

It will be recalled that only the Secretary of State has the power to refer merger cases. In addition, in the event of an adverse public-interest finding, the MMC's duty is to make *recommendations* to the Secretary of State. This politicisation of merger policy and the concentration of power in the hands of one politician has been criticised in so far as it diminishes the authority of the Commission and encourages lobbying, a practice much in evidence in recent cases such as BTR/Pilkington, GEC/Plessey and Nestlé/Rowntree. Of the two, the power to refer and the power to decide, the former is probably the more important. Certainly in practice there have been several occasions when the Secretary of State has not taken the Director General's advice on a reference but only once has the majority verdict of the MMC been rejected. If a merger bid is referred and the MMC's recommendations are not accepted there is at least a great deal of information available on which to judge the issues. However, if the Secretary of State refuses to make a reference there may be little information available to judge the wisdom of his decision. The case for placing the power to make a reference firmly in the hands of an independent body such as the OFT seems to be a compelling one. The issue of the power to decide is less clear-cut. Certainly it must be more difficult for the Secretary of State to ignore the recommendations of the MMC than it is for him to refuse to make a reference in the first place. Furthermore, there is the argument that it is right that the ultimate decision whether the public interest should prevail over private rights should be taken by a minister answerable to Parliament.[20]

Since 1984, successive Secretaries of State have stated that references are made primarily on competition grounds. However, once a reference is made there is no guarantee that the Commission will give competition issues the same prominence. For this the culprit is s. 84 of the

---

[20] Le Quesne, Sir G., (1988) 'The Monopolies Commission at Work' in George, K. D. (ed) *Macmillan's Mergers and Acquisition Yearbook*, 1988. (Sir Godfrey Le Quesne was chairman of the Monopolies and Mergers Commission 1975-1987).

1973 Act. As Le Quesne has pointed out: 'in practice there are certain questions which figure prominently in practically every inquiry, the foremost being competition; but the language of s. 84 means that there is no limit to the matters which may have to be considered in relation to the public interest'.[21] These matters include regional issues and the balance of payments effects of mergers, considerations which do not fit happily into competition policy.

The case for reversing the burden of proof was considered in 1978 by the Green Paper 'A Review of Monopolies and Mergers Policy', and rejected in favour of a neutral position, but no action was taken on this recommendation. It was argued that reversing the burden of proof would result in the MMC being swamped with referrals. The possibility of reversing the presumption for certain types of merger that give rise to particular concern, for instance those that create a large market share or involve very large companies, was also rejected. It was argued that reversing the burden of proof for this sub-category 'would be a strong deterrent against bringing forward merger proposals above the threshold even though they might be desirable in the national interest'. (para. 5.16)

The 1988 review of mergers policy also decided to leave the onus of proof unchanged. The defence of existing policy outlined in the review (paras. 2.8-2.13) is that given the presumption that the vast majority of mergers raise no competition or other objections and are best left for the market to decide, 'it would be inconsistent to reverse the burden of proof and to require those proposing a merger to demonstrate that this proposal would be positively in the public interest. This would make takeovers much harder to carry out, and would have a damaging effect on efficiency by weakening the discipline of the market over incumbent company managements'.

These arguments are not convincing. Even with the onus of proof reversed the vast majority of mergers would, as at present, be cleared quickly in the initial screening phase. Only those which, in the opinion of the OFT, presented a threat to competition, would be referred to the MMC. The severer treatment of merger proposals above some sensibly defined threshold is exactly what should be aimed at. The 'severer treatment' would need to come both in the preliminary vetting

---

[21] Le Quesne, G. *op. cit.*

procedures of the OFT and in the reversal of the burden of proof for those cases that were referred. If we are not prepared to accept that, *a priori*, the greatest danger to competitive behaviour comes from the most highly concentrated industries there is not much of a basis left for policy. As to *deterring* large mergers, that again is a desirable outcome. There would be a strong disincentive to companies to come forward with merger proposals which rested on filmsy arguments. However it would not deter merger proposals being made where the parties concerned felt that they had a convincing case. This is as it should be.

### 4.5. Concluding comments

In spite of much criticism over the years UK merger policy has changed little since its inception. The 1988 review made proposals for streamlining procedures but the fundamental weaknesses of the policy remain. In particular a more effective policy would seem to require the following:

(a) A mandatory rather than a voluntary pre-notification procedure for all mergers above a given size. The argument presented in the 1988 Mergers Review that this would impose an unacceptable administrative burden on business is not persuasive. A mandatory system would have the merit of ensuring that the OFT is fully informed of all important merger activity before the event.[22]

(b) The OFT should have the power to decide on all references where, in its view, important competition issues are involved, with the Secretary of State's powers in this respect being limited to those cases where the main issues at stake are other than competitive ones.

(c) The public-interest criterion should be changed to focus more sharply on the consequences of a merger for competition. Efficiency benefits should remain a permissible defence but the onus of proof should be reversed with the parties to a merger having to demonstrate, to the satisfaction of the MMC, that the merger would be in the public interest.[23]

---

[22] On this and other proposals for reform see: Hay D. and Vickers J (August, 1988): 'The Reform of UK Competition Policy', *National Institute Economic Review*.
[23] See Fairburn, J. A. and Kay, J. A. (eds) *op. cit* for a recent survey of mergers and merger policy in the UK.

Needless to say, the foregoing argument is not intended to deny that takeover, or the threat of takeover, often has a beneficial effect on managerial efficiency. But it is necessary to emphasise that mergers may also have important effects in creating or strengthening monopoly positions, and where this danger is greatest it should be for firms to demonstrate that any such detriment is outweighed by efficiency benefits.

## 5. VERTICAL RELATIONSHIPS

Vertical relationships are dealt with under all of the four main competition Acts. They can be investigated as part of a full-scale monopoly investigation under the 1973 Fair Trading Act; some, such as franchising agreements and exclusive dealerships fall under the 1976 Restrictive Trade Practices Act; resale price maintenance is caught by the 1976 Resale Prices Act; and, in so far as they might involve anticompetitive practices, vertical relationships may also be investigated under the 1980 Competition Act.

Many anticompetitive practices are in fact related to vertical arrangements between firms—they include ties, line-forcing, exclusive dealing, discriminating discounts, and refusal to supply. These, together with resale price maintenance are considered in this section.

### 5.1. Ties

A tie-in sale occurs where the supply of one good or service is tied to that of another. For instance, the supplier of a copying machine (the tying good) may require that supplies of photocopying paper (the tied good) are also purchased from him. A tie also exists where a customer cannot purchase a single item in a supplier's line of goods without buying a line or a full line. This practice is known as line-forcing or full-line forcing.

There are several ways in which ties of this sort may be brought about. The tied arrangement may be expressly stated as a condition of supply, or it may be induced by special pricing arrangements or discounts which make it more advantageous for a customer to purchase two or more products from the same supplier than to purchase them singly from different sources. Tied sales may also occur because of the absence of alternative sources of supply for the tied good or service, as may occur, for instance, in the servicing of technologically-new products.

Tied arrangements may be motivated by a variety of reasons including:

(a) the preservation or extension of market power,
(b) as a means of practicing price discrimination,
(c) to ensure the proper functioning of the tying product,
(d) in order to reap economies of scale or of scope and
(e) as a means whereby a regulated body may be able to avoid price controls.

This list of possible motives suggests that tying arrangements are unlikely to be universally beneficial or harmful, a conclusion that was reached by the MMC's 1981 report on *Full-line Forcing and Tie-in Sales*. That report recommended a-case-by case approach and also suggested guidelines for assessing the likely consequences of tying arrangements. These were:

(a) that tie-in arrangements are unlikely to have serious anti-competitive effects unless the supplier has market power in the supply of the tying good or service;
(b) that the seriousness of any anticompetitive effect will also depend on the structure of the market for the tied good, and in particular the proportion of the market foreclosed by the tie and the number of alternative suppliers;
(c) that ties such as those relating to consumable materials, spare parts and servicing may be anticompetitive but may be defended on technical grounds;
(d) that the anticompetitive effects of a tie are likely to be more serious if they are associated with an insistence on exclusive dealing.

Perhaps the best known example of tie-in arrangements is found in the supply of beer. Most brewing companies in Britain wholesale and retail their beer, and as retailers they own about 75 percent of public houses. Some of these are managed houses in which the publican is an employee of the brewing company, others are tenanted, where the publican pays the brewer a rent for the premises. In both categories the brewer specifies what beers may be sold and where they must be bought (usually from the brewer himself). In addition, about half the 'free houses' are tied to a brewer by a loan. In these cases it is common for the publican, in return for a low interest loan, to agree to take a certain quantity of a range of beers from the brewer. In practice is is also common for tenants to take a high proportion of non-beear drinks from their landlord.

The MMC found that these tie-in arrangements formed the basis of a complex monopoly (see p. 118). The Commission recognised that vertical integration can have beneficial effects on both allocative and managerial efficiency, but that in the circumstances of the brewing industry it has also led to ineffiency and higher costs. In particular the captive markets of the major brewers reduces the effectiveness with which capacity adjustments are made resulting, for instance, in unexploited economies of scale. Vertical integration can also have anticompetitive effects; by restricting competition at each stage in the supply of beer the vertical arrangements of the brewers had resulted in higher prices and a restriction in consumer choice.

In view of the powerful market position of the major brewers (the 6 national brewers accounting for 75 percent of beer production, 74 percent of brewer-owned public houses and 86 percent of loan ties) and the fact that their market position had been strengthening, the Commission concluded that basic structural changes were needed to remedy the situation. Its most radical recommendations were that no brewer should be allowed to own more than 2000 public houses and that loan ties should be abolished. Following intense political lobbying by the brewers the Secretary of State decided not to accept these recommendations. He did, however, decide to introduce measures, some of them recommended by the Commission, to loosen the ties. Instead of forced divestment of public houses it is intended that brewers owning more than 2000 pubs will have to keep at least 50 percent of the additional pubs as free houses, leased free of ties to the company's products. And instead of abolishing loan ties the intention is to make it easier for publicans of free houses to extricate themselves from a tie and to switch from one brewer to another. Whether these less radical measures will be effective remains to be seen.[24]

---

[24] For details of this interesting case study of vertical arrangements between firms see Monopolies and Mergers Commission, *The Supply of Beer*, HMSO, Cm 651, 1989.

## 5.2. Exclusive Dealing

Exclusive dealing too has to be judged on the merits of individual cases. The dealer benefits from having to face less competition so that a higher margin on sales can be maintained. The manufacturer hopes to benefit by having his product sold by enthusiastic dealers who, because of a higher margin than would otherwise be possible, have an incentive to carry larger stocks, offer better services and spend more on advertising and other sales promotion activities.

Consumers may gain or lose from these arrangements. For many products differentiation is demanded by consumers and in so far as it is strengthened by exclusive dealing this may, up to a point, be beneficial, although this result may be less likely if it is achieved by competitive advertising expenditure than by competition in providing good after-sales service. Generally, exclusive dealing will blunten competition within brands but this may not be of much concern if competition between brands remains vigorous.

Exclusive dealing is unlikely, therefore, to be invariably harmful or invariably beneficial to consumers. Beneficial effects are most likely where it aids entry of a new competitor, or where the product market is unconcentrated and differentiation is weak so that strong inter-brand competition can exist. It is most likely to be harmful when the product market is highly concentrated and dominant suppliers can use exclusive-dealing arrangements to foreclose a large proportion of the market from other competitors.

Considerations such as these are reflected in those Commission reports where exclusive dealing was an issue. For instance in *Ice Cream and Water Ices* (1979) exclusive dealing involving unfranchised outlets was found to be against the public interest because it restricted competition between the two main suppliers who together accounted for over 60 percent of supplies, and also because it reduced opportunities for other suppliers. These detriments were judged to outweigh any advantage which exclusive dealing might have had in respect of economies in distribution. In *Frozen Foodstuffs* (1976) exclusive dealing by Unilever (33 percent market share) was not found to be against the public interest because the existence of other large suppliers with similar arrangements indicated that their ability to compete was not being impaired. The practice was also found not to be an important obstacle to smaller suppliers and new entrants.

## 5.3. Franchising

Exclusive-dealing arrangements are frequently associated with franchising where the franchisor requires the franchisee to purchase goods exclusively from him. In many cases this undoubtedly is a reasonable requirement. For instance the goods sold may be an essential part of the franchise, being a unique product supplied by the franchisor and bearing his distinctive trade mark. A good example is the fast-food market with franchise names such as Wimpy, and MacDonalds. In this market there is keen competition between close substitutes and it cannot be said that franchising is anticompetitive in any meaningful sense.

In other cases, however, the position is likely to be less clear-cut, particularly where the obligation is to obtain goods from the franchisor which could readily and without disadvantage be obtained elsewhere, or where the goods cannot be regarded as an essential part of the franchise package. Whether goods can be so obtained and whether or not they are an essential part of a franchise package is a matter of some debate as is illustrated in the Commission's report on *Car Parts* (1982).

This report examined the practice whereby suppliers required persons to whom they supplied car parts to acquire those parts exclusively from them or from sources approved by them. The main arguments put in defence of the practice were:

(a) the sale of cars and of spare parts and the servicing of vehicles is a package which a car-supplier affords the customer and which he has to offer in order to remain competitive;
(b) the sale of unsuitable or unsafe parts would have a damaging effect on the car-maker's reputation and insistence on exclusive buying is a safeguard against the use of such parts;
(c) exclusivity gives the consumer a clear-cut choice between obtaining 'genuine' parts from a franchised outlet and 'spurious' parts from other outlets.

The Commission did not accept these arguments. It pointed out that a requirement that a franchisee carry a certain minimum stock of parts obtained from or approved by the franchisor would be sufficient to ensure the availability of such parts to the customer without the added insistence that the franchisee obtain no competing parts from any other source. Furthermore, franchisees had their own reputations to safeguard and were unlikely to jeopardise it by selling and fitting inferior or unsafe parts. Car-component suppliers argued that they were also anxious to preserve their reputations and pointed out that they produced parts for the replacement market that were identical with, or equal in quality to, those supplied by car manufacturers and importers to their franchised dealers. Particularly telling was the fact that the major UK car manufacturers themselves also supplied parts for other makes of car under their 'all-makes' marketing programmes and often obtained these parts from the component manufacturers.

The Commission concluded that exclusive buying restricted competition in the supply of car parts and since there were no compensating benefits the practice was against the public interest.

### 5.4. Resale Price Maintenance and Discounts to Retailers

UK competition policy has been more severe towards RPM than to any other aspect of business behaviour. The 1964 Resale Prices Act which introduced the general prohibition of individual RPM provided for exemption in cases where its abandonment would harm consumers' welfare: by reducing the quality or variety of goods or the number of retail outlets; by increasing retail prices; by allowing goods to be sold under conditions dangerous to health; or by reducing necessary services provided with or after the sale of the goods. However the narrowness of these gateways compared to those that apply to other forms of restrictive agreements has made it difficult for manufacturers to defend the practice. In only two cases—pharmaceutical products and books—has there been a successful defence, and these are the only products on which RPM can legally be enforced.

UK legislation does not prevent suppliers from fixing maximum resale prices, neither does it prevent them from recommending a price at which their goods should be sold. There is an obvious danger that the practice of recommending resale prices might have the same result as RPM. This possibility was recognised in the MMC's 1969 report *Recommended Resale Prices*, but no evidence was found that the practice was invariably harmful.

The main arguments which are usually put forward in favour of RPM are that: it protects retail margins with the result that the retailer can offer better service to consumers, which indirectly benefits the manufacturers; and it prevents the use of a product as a loss-leader, a practice which can injure both the retailers who specialise in the loss-leader product and the reputation of the manufacturers of the product. It will be noted that these arguments are similar to those that apply to non-price vertical restraints such as exclusive dealing.

Critics of RPM argue that it denies the consumer the choice of price/service combinations; it is an obstacle to the development of more efficient forms of retailing; it results in excess capacity in the distributive trade, and tends to bolster manufacturers' price agreements by making it easier to detect cheating.[25]

One of the likely consequences of the tough official stance against RPM is that it contributed to the trend towards higher concentration in the retail trades and to the emergence of the big multiple chains. These chains have been able to use their buying power to extract larger discounts from suppliers which has contributed further to the increase in concentration. As a result the major area of concern is quite different now to what it was in the 1950s and 60s. Then it was a concern about obstacles to the emergence of more efficient forms of retailing; now it is about the bargaining power that the new forms of retailing have acquired.

In the 1981 report *Discounts to Retailers*; the MMC concluded that on the suppliers' side discriminatory discounts were more often than not a reflection of effective competition between suppliers confronted by powerful buyers rather than a reflection of monopolistic behaviour. On the retailing side the practice had contributed to increased efficiency by helping to break down traditional margins, and in general was a reflection of a highly competitive situation, which had benefitted the consumer in the form of lower prices. The report concluded that no overall measure of prohibition or regulation of discriminatory discounting was necessary or desirable but that concentration in the retail trades could go too far to the disadvantage of both suppliers and consumers.

### 5.5. Refusal to Supply

Refusal to supply may be a thing done to exploit a monopoly position as was found to be the case in the MMC's investigation of British Gas

---

[25] For an early discussion of the issues see Yamey B. S. (ed) (1966) *Resale Price Maintenance*, Weidenfeld and Nicolson, London.

(1988). On the other hand it may simply be a reflection of the method of distribution which a firm considers to be in its own best interests, with no significant anticompetitive implications.

The first refusal to supply case investigated under the Competition Act 1980 was that of *Bicycles* (1981). The matter under investigation was Raleigh's policy of refusing to supply certain large-discount stores. As compared to the traditional specialist retailers the discount stores were high-volume low-margin outlets offering less service. Raleigh argued that its distribution policy was in its commercial interests because, amongst other things, it maintained Raleigh's brand image which had been built up hand in hand with the goodwill of its dealers. It was important for Raleigh's competitive position that the traditional outlets were protected.

The Commission argued that Raleigh's policy of witholding supplies from some discount stores restricted competition in the retailing of Raleigh bicycles and that the restriction was material, because the company was the leading supplier of bicycles in the UK and because its refusal was directed against a particular class of retailer—namely discount stores. The Commission thus concluded that Raleigh was guilty of an anticompetitive practice and since it was unable to discover benefits sufficiently important to outweigh the detriments it concluded that Raleigh's policy was against the public interest.

It has been suggested that the Commission's reasoning could be used as a precedent for the position that *any* refusal to supply is anticompetitive.[26] This is not so. The Commission considered the restriction to be material in part at least because of Raleigh's position as the largest UK supplier. It considered the argument advanced by Raleigh that a supplier had the right to determine how its goods should be distributed. However, the Commission could 'not accept that this is an *entirely* unfettered right that can be upheld in *all circumstances... The case for some restriction of the right is particularly strong where a manufacturer is a dominant supplier and his goods are differentiated, or are perceived by distributors and the public to be differentiated from other manufacturer's goods of the same kind*' (emphasis added). There is no doubt, and there was no disputing the fact, that Raleigh was refusing to supply known price cutters, that the Company was fearful that

---

[26] Kay, J. A. and Sharpe, T. A. E., (Nov., 1982) 'The Anti-Competitive Practice' *Fiscal Studies*.

supplying these outlets would lead to the loss of many traditional outlets and that the Company was the dominant supplier of a differentiated product.

The proper criticism that may be made of the Commission's verdict in this case is that it was based on a degree of market power that did not exist. Had Raleigh's market share carried with it substantial market power and the ability to dictate retail prices the verdict would have been appropriate. However, this was not the case. Raleigh's market share had fallen from 67 percent in 1972 to 40 percent in 1980 while imports had increased from 6 percent to 36 percent over the same period. Competition at the retail end of the market was vigorous. Indeed the Commission recognised that Raleigh's refusal to supply might have contributed to its decline. If it is conceded that Raleigh's refusal to supply was more a cause of its decline in competitiveness than a manifestation of monopoly power then its behaviour could hardly be described as anticompetitive.

### 5.6. Concluding Comments

Vertical restrictions come in many different forms and with one exception the approach in the UK has been 'open-minded'. The exception is price restraint where the policy towards RPM has been close to *per se* prohibition.

For non-price restraints both the general reports and the industry investigations of the MMC recognise that such restraints can have efficiency-enhancing effects. The argument for such benefits however is most likely to find favour where there is effective horizontal competition both at the supplier and the retail end of the market and where the restraints do not raise entry barriers. In other words, restraints on intra-brand competition are of little concern if there is effective inter-brand competition.

Many economists have questioned the wisdom of this sharp distinction between price and non-price vertical restraints.[27] Both RPM and the allocation of local markets through exclusive-dealing arrangements may serve the same end of protecting distributors, who have invested in establishing a reputation, from the price-cutting behaviour of free-

---

[27] See Hay, G., (1985) 'Vertical Restraints', *Fiscal Studies*, vol. 6.

riders. On the other hand, if primacy is to be given to maintaining flexible market conditions then support should not be given to practices which may form the basis of collective action by manufacturers or coercion by retailers. From this point of view price restraints are probably more dangerous.

Be that as it may, future policy towards vertical restrictions in the UK is at present somewhat uncertain. As already noted in Section 2 the Government has in mind a policy that will prohibit anticompetitive agreements—vertical as well as horizontal, with an exemption procedure including the possibility of block exemptions. No explicit consideration is given in the 1988 review, however, to resale price maintenance although it may be inferred from the intention to prohibit all price-fixing agreements that the hard-line approach to vertical price restraints will continue.

# French Competition Policy in Perspective

F. JENNY

Professor of Economics, E.S.S.E.C. Rapporteur General, Conseil de la Concurrence

## 1. THE EMERGENCE OF COMPETITION LAW IN FRANCE

Until quite recently, it was widely assumed in international circles that anticompetitive pratices were widespread in France and that French authorities were largely unconcerned with promoting competition. This view was fostered by the well-known, long-standing French tradition of government intervention in market mechanisms ranging from price controls to subsidization of key industries and centrally planned restructuring of firms, as well as by an obvious lack of a competitive spirit of French firms dealing in international markets.

Thus the claim that the enactment of the 1986 ordinance on 'price freedom and competition' was a major breakthrough for the promotion of the competitive free market ideal in France, was widely accepted.

To a large extent, the idea that French public-policy-makers did not care about competition policy was justified until the mid-1970s and although antitrust legislation had been enacted in 1953, its implementation was at first seen by academicians and the business community as little more than window-dressing.

However, since the mid-1960s important changes have occurred in the attitudes of public-policy-makers, the business community, and consumers with respect to the benefits to be expected from the development of competition. In addition, there have been equally important changes in the nature and the vigor of enforcement of antitrust legislation.

## 1.1. The context of French economic policy in the immediate post-war period.

Four aspects of French economic policy are important to understand what characterized competition policy during the '50s, the '60s and the early '70s.

First, the need to reconstruct the French economy after World War II and the existence of powerful business organizations led to a particular type of government intervention in the market mechanisms exemplified by the French indicative planning process. Business community leaders and public officials jointly monitored the economy.

Second, concern about black-marketeering during the immediate post-war period of reconstruction and about inflation during the subsequent period of rapid economic development led public authorities to rely heavily on price controls.

Third, concern with developing the French industrial potential and an inclination toward engineering achievements reinforced the commonly held idea that distributors did not contribute significantly to the creation of real wealth but were mainly bothersome intermediaries.

Fourth, the progressive involvement of France in the European Economic Community at the beginning of the '60s led to the belief that France's main disadvantage in the realm of international competition lay not in an inability to innovate, but was a result of weaknesses in its industrial structure. It was widely held that exports expansion would occur naturally if French firms could benefit from economies of scale in production. Increasing the size of French plants and French firms seemed the obvious way to rapidly achieve the desired result. This view, which was popularized in 1967 by J. J. Servan Shreiber in his book 'The American Challenge'[1], led to a merger wave between 1965 and 1975. A policy granting financial aids to firms willing to engage in external growth was supported by public officials in part because they thought that it would be easier to monitor a small number of large firms in each industry than a large number of small firms. It was also supported by the business community which realized that concentration that might possibly lead to a decrease in unit costs of production

---

[1] 'Le défi américain', Jean Jacques Servan Schreiber, Denoel, Paris, 1967.

## 1.2. The 1953 antitrust statutes

The enactment of the first French competition antitrust statute (decrees of August 9, 1953 and January 27, 1954) was basically at odds with the overall economic approach of public-policy-makers. At the time, the enactment of a competition law was mainly seen as a complement to the price control ordinance adopted in 1945 (Ordonnance No. 45-1483 du 30 juin, 1945, relative aux prix) and the decree of 1953 was an amendment to the 1945 ordinance. Indeed, due to a lack of understanding of economic theory, it was generally believed that anticompetitive behavior should be opposed because it led to higher price levels and therefore contributed to inflation, rather than because competition would lead to a more efficient allocation of resources.

The major characteristics of the 1953 competition statutes can be explained by the aforementioned factors.

The newly-added article, 59 *bis* of the ordinance of 1945 on prices, stated that: 'are unlawful, subject to the provisions of article 59 *ter*, all concerted actions, conventions, express or tacit agreements, or coalitions whatever their form or their cause, if their object or their potential effect is to impair the full exercise of competition by preventing a decrease in costs or in selling prices or by facilitating an artificial increase in prices'.

The text of article 59 *bis* clearly indicated the government preoccupation with price movements.

Article 59 *ter* stated that 'are not subject to the provisions of article 59 *bis* concerted actions, conventions or agreements: 1) if they result from a law or a decree; 2) if their authors can justify that their effect is to ameliorate or increase production or to ensure the development of economic progress through rationalization and specialization'.

Thus it was assumed that competition was not necessarily the most appropriate way to increase efficiency or production and that concern about competition should not interfere with the industrial policy favored by business leaders and supported by the French government.

The preliminary investigations of potentially anticompetitive agreements were to be conducted by the Price Division of the Ministry in charge of Economic Affairs. Upon completion of those investigations, the Minister could refer the case to an administrative body (Commission Technique des Ententes) which was to give an opinion as to whether or not the examined practices were prohibited. If the Commission technique des Ententes considered that there was a violation of the competition statutes, the Minister could either send the firms or organizations involved an official letter asking them to discontinue the practice or he could refer the case to the courts.

The 1953 decree did not establish merger control or a control of abuse of market power by individual firms having a dominant position on a market.

The 1953 provisions were modified on several occasions during subsequent years. A review of those changes reveals a slowly growing sophistication with respect to the analysis of market mechanisms.

In 1963, ten years after the enactment of the original provisions against cartel agreements had been enacted, a provision prohibiting anticompetitive practices by firms having a dominant position on a market or a monopoly was added to article 59 *bis* of the 1945 ordinance on prices (law No. 63-628 of July 2, 1963).

In 1967, article 2 of the ordinance No. 67-835 of September 28, 1967 amended article 59 *bis* of the 1945 price ordinance prohibiting cartel agreements in a way that can be interpreted as a slight move away from the idea that competition was primarily designed to prevent price increases. The revised text of article 59 *bis* stated that 'are unlawful, subject to the provisions of article 59 *ter*, all concerted actions, conventions, express or tacit agreements, or coalition whatever their form or their cause, if their object or their potential effect is to prevent, restrain or alter competition; for example by preventing a decrease in costs or in selling prices, by facilitating an artificial increase in prices, by preventing technical progress, or by limiting the ability of other firms to compete freely'.

The ordinance No. 67-835 of September 28, 1967 also amended article 59 *ter* which now read: 'are not subject to the provisions of article 59 *bis* concerted actions, conventions or agreements: 1) if they result from a law or a decree; 2) if their authors can justify that their effect is to ensure the development of economic progress, in particular by an increase in productivity' (rather than the previous phrasing which stated: 'if their effect is to ameliorate or increase production or to ensure the development of economic progress through rationalization and specialisation').

However, these changes were of marginal importance and the promotion of competition on the market-place did not figure high on the agenda of government authorities in charge of economic matters until 1977. Besides the obvious unwillingness of public authorities to rely on market forces in a competitive environment to achieve efficiency, the system designed in 1953 was largely ineffective because the legal means of the Minister in charge of Economic Affairs were insufficient to dissuade firms from engaging in anticompetitive practices either because they did not care about the warnings that he could issue or because the courts did not handle speedily and effectively the cases which he decided to refer to them.

**1.3. The 1977 reform**

The change that occured in 1977 under the premiership of Mr Raymond Barre were decidely more significant. Three main factors played a role in the 1977 reform.

First, a personal factor. Mr Barre is an economist by training and therefore familiar with price theory. Prior to becoming Prime Minister, he had been chairman of the EEC Commission, which has always actively promoted competition at the European level.

Second, an economic factor. At the beginning of the '70s, research in the field of industrial organization revealed three major facts that challenged the wisdom of previous policy in the area of industrial structures: concentration on the market place had increased significantly during the previous decade partly, but not exclusively, due to the merger policy actively pursued by public authorities; the profit rates of firms in concentrated industries were significantly higher than the profit rates of firms in industries that had a more competitive structure; in many industries the largest firms were less efficient than their smaller competitors. These findings, combined with obvious examples of industrial policy failures in industries in which the government had tried to increase concentration (such as in the steel and computer industries) and the slowly emerging feeling that the lack of aggressiveness of French firms on international markets was partly a consequence of weak competition on their domestic market, led a vocal minority of economists and public-policy-makers to call for a strengthening of French competition law.

Third, an institutional factor. At the 1972 'Paris Conference', the heads of the EEC countries had agreed on the necessity of establishing a merger-control law at the European level. This plan was considered a threat by the segment of French bureaucracy in charge of industrial policy, which was still busily involved in promoting mergers among French firms. Indeed, French bureaucrats realized that if this merger control were to come into being, merger control at the EEC level would not necessarily be limited to mergers between firms of different countries. It could well mean that a merger between two large French firms holding a significant share of the European market (such as a merger between two large French automobile manufacturers), even one backed by the French government and considered essential from the point of view of domestic industrial policy, could in the future be challenged by 'faceless Eurocrats' on the ground that it might impair competition and trade within the EEC. Thus, although public-policy-makers in charge of French industrial policy generally still believed in the positive effect of concentration and mergers, they were led to favor the adoption of some kind of merger control at the French level (as long as that control was exercised by the government rather than by the courts) as a way to defeat the European merger-control project. Indeed they reasoned that it would be far easier to argue that the establishment of a merger control at the European level was useless, if they could show that French authorities were trying to curb potentially anticompetitive mergers at the domestic level.

Thus, for the all of the above, somewhat contradictory, reasons a major revision of French competition law took place in 1977. Four main provisions of the new law (Loi No. 77-806 du 19 juillet, 1977 concerning the control of economic concentration and the repression of illicit agreements and abuse of dominant positions) are worthy of comment.

First, the Minister in charge of Economic Affairs could henceforth impose administrative fines on firms found by the Commission de la concurrence (the new advisory body in charge of investigating cartel agreements and the behavior of firms holding a dominant position on a market) to have engaged in anticompetitive practices. Fines imposed by the Minister could not exceed amounts recommended by the Commission nor could they exceed 5% of the total annual sales of the firms involved.

Second, the Commission de la concurrence was established as an administrative body independent of the Minister of economic affairs. Cases involving potentially anticompetitive practices could henceforth be referred to the Commission by parties other than the Minister in charge of economic affairs, such as trade associations, consumer organizations and local governments.

Third, merger control was established. The control could occur in the case of horizontal mergers, if the market share of the merging firms exceeded 40% on any market 'at the national level', and, in the case of non-horizontal mergers, if the market shares of two of the merging firms exceeded 25% on different markets. The 1977 law stipulated that if the Commission de la concurrence found that the disadvantage for competition of the merger outweighed its contribution to economic and social progress, the Minister in charge of economic affairs, and the Minister in charge of the economic sector considered could prohibit the merger or order divestiture. However, the ministers decisions could not go beyond the Commission de la concurrence recommendations.

Although, the 1977 law did not change the substance of French competition law regarding the definition of anticompetitive agreements and practices of dominant positions, it was an important signal to the business community that public authorities were serious about discovering and curbing anticompetitive behavior.

Between 1977 and 1986, the Commission de la concurrence investigated 145 cases of potentially anticompetitive agreements (ranging from price fixing to market sharing, illicit exchanges of information among firms, attempts to block entry into an industry, selective or exclusive distribution contracts designed to prevent competition at the retail level, conspiracies in procurement markets etc . . .) or abuses of dominant positions.

Despite this, the 1977 law was, to a certain extent, considered a half measure. Indeed, although it was meant to affirm the commitment of the French public authorities to promoting competitive market mechanisms, the Minister in charge of economic affairs retained both preliminary investigative powers and the power to make final decisions about the cases. As it was clear that a Minister's decision could be influenced by political considerations, the effort to promote competition was often considered as just another tool added to the already considerable means for government interference in market mechanisms rather than as a way for the government to permit decentralized economic forces to run their course. This impression was reinforced by continued widespread recourse to price controls (until 1983) which enabled the Minister in charge of economic affairs to impose ceilings on yearly price increases in some industries.

Nevertheless, French firms slowly but surely recognized that if they engaged in anticompetitive practices, they ran the risk of being caught and penalized. The Commission de la concurrence did not hesitate to recommend significant fines in cases in which firms or their trade associations were found to have flagrantly violated the law (up to 1 million dollars for some firms). The Minister of Economic Affairs concurred with the analysis of the Commission in most cases, although in some instances he reduced the fines proposed by the Commission (frequently between 1977 and 1983; more rarely afterwards). It was the imposition of fines in certain cases, more than orders to discontinue prohibited practices, which played a crucial role in popularizing the notion that the promotion of competitive market processes was a worthwhile entreprise. This occurred because monetary sanctions were considered highly newsworthy by the press and received wide coverage. Thus the idea that firms could individually or collectively abuse their market power and that consumers had a collective interest in the prevention of such practices slowly but surely seeped into the French conventional wisdom.

It is interesting to note that the change that occurred in 1981, when a socialist government replaced the conservative governments which had been in power since 1958, did not slow down the trend towards a more vigorous enforcement of the competition. Indeed, in 1981/1982, although the Socialists did not favor the free-market ideal, they were suspicious of economic power and considered competition law a useful tool to curb 'abuses' of economically powerful firms. What is more, after 1983, as the 1986 legislative elections were drawing to a close, they chose to undercut the conservative opposition by adopting a free market approach.

### 1.4. Other restraints on business practices

In addition to the prohibition of anticompetitive cartel agreements and of abuse of market power by dominant firms, a number of other business practices were prohibited during the post-war period, namely: resale price maintenance, refusal to deal, price discrimination, and reselling at a loss. The prohibition, regarding these practices were until very recently, *per se* prohibitions. However, these prohibitions were not enforced by the Commission de la concurrence but were handled through the French civil court system or through an administrative procedure.

During the early '50s, the development of large-scale retailers and of discount stores was hampered by the fact that traditional (small-scale) distributors threatened to boycott, and in some cases actually boycotted, manufacturers who did not maintain resale prices. Such boycotts were not necessarily indicative of the existence of illegal cartels among the retailers involved but reflected the corporatist tendency of a large majority of the distributors.

Partly because it believed that small scale distributors were often inefficient and partly because it was under the mistaken impression that the development of large scale distributors and discount stores would reduce inflation, the French administration tried to remedy the situation at the end of the '50s by making it illegal for manufacturers to refuse to sell or to impose resale prices.

During the '60s, traditional distributors retaliated by alleging that manufacturers discriminated against them and in favor of large-scale distributors and that those distributors were using unfair methods of competition by reselling certain products at a loss (to attract consumers) while making a profit on others. Using the considerable political clout that they derived from their large number, traditional distributors were successful in lobbying for legislation making it illegal *per se* to ask for or to grant a discriminatory advantage or to resell at a loss.

In 1973, traditional distributors scored another victory in their fight against the development of large-scale low-margin retailers when the 'Loi ROYER' subjected the creation of new retail stores above a certain size to the prior authorization of local administrative bodies called 'Commissions d'Urbanisme Commercial' (CUC) in which local traditional distributors were represented and could make themselves heard.

However, for various reasons, these proved to be illusory victories for the traditional distributors.

The *per se* ban on price discrimination ceased to be enforced by public authorities around the middle of the '70s not only because in many cases it was often difficult to prove the existence of such discrimination but also because the government slowly but surely realized that in most cases price discrimination enhanced competition rather than hampered it.

The *per se* prohibition on refusal to deal was easier to enforce and, by and large, manufacturers were forced to sell their products to large-scale distributors even when this displeased traditional retailers. However, numerous manufacturers, eager to avoid a strong negative reaction from their traditional small-scale retailers, tended to give a substantial part of the discounts that their large-scale distributors were entitled to in the form of conditional discounts (i.e., discounts paid at the end of each year and based on the total volume purchased during the year by the distributor). As the price below which distributors were prohibited from reselling (by the *per se* prohibition on reselling at a loss) was the invoice price and not the actual purchase price (invoice price minus year-end conditional discount), in some cases the effect of this practice was to force 'discounters' to sell at prices higher than those they would have chosen spontaneously and therefore to somewhat limit price competition among different types of distributors of the same good.

Whenever successful large-scale distributors found it difficult to obtain the necessary authorization (from the CUC) to open new stores in areas where they wanted to expand, they tended to use their profits to buy existing stores (as no prior authorization was necessary in such cases). Thus concentration in the retail sector increased significantly, partially as a result of the very laws which had been designed to protect small traditional distributors.

The development of concentration in the retail sector led to an increase in the buying power of some distributors who were in turn able to threaten manufacturers with refusal to buy if they did not offer them additional (discriminatory) rebates. Manufacturers reacted by giving in to the demands of the most powerful or the most determined nationwide large-scale retailers. As a result, price discrimination by manufacturers became widespread and buying conditions offered to traditional distributors and their large competitors became more and more unequal.

The slowdown of economic growth at the beginning of the '80s intensified competition among nationwide large-scale retail chains and led each of them to intensify its efforts to lower resale prices and to secretly obtain ever more favorable buying terms from its suppliers.

Furthermore, competing distributors began to set up common buying agencies ('supercentrales d'achat') in order to increase their leverage over their suppliers. Thus, concentration on the buyer side increased even further.

Thus, during the '60s and the '70s significant changes occurred in France regarding the economic relationship between manufacturers and distributors of consumer goods. Large-scale distributors have become a force to reckon with both for traditional distributors and for manufacturers.

At the beginning of the '80s, the complex and burdensome legislation which accrued from successive attempts by the government to monitor the changes in the relationship between manufacturers and distributors was widely denounced as uneconomical, ineffective and unjust. Manufacturers of brand goods, in particular, strongly denounced the *per se* prohibition of refusal to deal, arguing that this prohibition put them at the mercy of powerful distributors which could force them to grant illegal discriminatory price concessions under the threat of boycott since they were not subject to a parallel prohibition of refusal to buy.

### 1.5. The 1986 ordinance on price freedom and competition

The aforementioned considerations help us to understand the rationale behind most of the important features of the new competition ordinance of 1986 (Ordonnance No. 86-1243 du ier decembre, 1986 relative à la liberté des prix et de la concurrence).

To underline its commitment to the concept of a free-market economy, the new administration repealed the unpopular Price control ordinance of 1945. Article 1 of the 1986 ordinance 'on the freedom of prices and competition' states that: 'prices are freely determined through competitive process'. However, the government retained the power to control the prices of a small number of predetermined products or services, mostly in the health or transportation sectors, for which price competition is limited either because of the existence of a natural or legal monopoly or because of an acute shortage.

In addition, the government relinquished its power of decision regarding anticompetitive cartel agreements and abuses of market power. A quasi-judicial body ('Conseil de la concurrence') charged with prosecuting such practices was created. Composed of 16 members (magistrates, economic and legal experts and members of the business community), this independent body has investigative powers and can impose fines or give injunctions to firms found guilty of having violated the law. The decisions of the Conseil de la concurrence can be appealed to the Paris Appeals Court. Alleged anticompetitive practices can be referred to the Conseil de la concurrence by firms as well as by trade organizations, unions, local governments or the Minister in charge of economic affairs.

The Conseil de la concurrence is also charged with advising the government, parliamentary commissions, business and consumer organizations, and unions on general questions related to competition.

As previously mentioned, in spite of the fact that the 1977 competition law established a possibility of controlling mergers and in spite of the poor performances of a number of firms that had merged in the '60s and the '70s, there is still a lingering feeling among French bureaucrats in charge of industrial policy that economic concentration is necessary to ensure the competitiveness of French firms and that the government should play an active role in the area of industrial structures. It is thus not surprising to observe that the government has not relinquished its decision-making power in the area of mergers.

The Minister in charge of economic affairs retains sole authority for referring mergers to the Conseil de la concurrence; after referral, the Conseil de la concurrence delivers an opinion as to whether the merger should be allowed (and if so under which conditions) or whether it should be prohibited. Whereas in the system established in 1977, the Minister in charge of economic affairs and the Minister in charge of the economic sector concerned could not go beyond the recommendation of the Commission de la concurrence, they are now free to follow or ignore the opinion of the Conseil de la concurrence.

As far as the substance of the law is concerned, the main provisions prohibiting cartel agreements and abuses of market power by firms holding a dominant position remained unchanged. However, the debate concerning the emerging power of distributors and their relationship with manufacturers led to some important changes in the French competition legislation. Three main questions were raised during the preparatory phase of the new competition ordinance:

— should antimerger law be revised to enable public authorities to control mergers in the distribution sector?

— should the practices of firms that do not have a dominant position on a market but which wield substantive bargaining power towards their suppliers or buyers be controlled?

— should existing legislation on vertical restraints of trade be revised?

### 1.5.1. Merger control

Until December 1985, the ability of the Minister in charge of economic affairs to control mergers was limited to two situations: in the case of horizontal mergers, the merging firms had to have a combined market share equal or superior to 40% of the domestic consumption of a product at the national level; when the merger was not horizontal, at least two of the merging firms had to have a market share of at least 25% of the market of a product at the national level.

In spite of the fact that the merger control law was not vigorously enforced, manufacturers consistently denounced it as unfair by arguing that the criteria used to determine which mergers could be controlled should not be the same for the distribution sector and for the manufacturing sector. Indeed, they pointed out that if in most cases manufacturers compete at the national level, competition among distributors is mostly local. Therefore they argued that a merger between two distributors could well limit competition even if it could not be controlled because it did not meet the market-share criterion written into the law. They further argued that although a large manufacturer might easily have a market share superior to 40% of the national market, not even the largest distributors, who wield considerable buying power (and occasionally have a dominant position at the local or regional level), account for 40% of the retail sales of any given category of goods at the national level.

The clamor of manufacturers convinced that merger legislation was unfair to them, was thus based on their fear of the growing concentration in the distribution system as well as on the fact that if merger legislation could be used to prevent manufacturers from gaining 'too much selling power', it could not, in fact, be used to prevent large distributors from gaining 'too much buying power'.

In December 1985 some changes making it easier to control mergers were introduced in the merger legislation. An amendment to the 1977 Competition Law (loi No. 85–1408 du 30 decembre, 1985) provided that thereafter any merger, whether horizontal or non-horizontal, in which the merging firms bought or sold more than 25% of a product at the national level or on a substantial part of the national market could be controlled if it was likely to decrease competition.

However, for two reasons the amendment to the 1977 law did not quite meet the demands of the manufacturers: first, it was obvious that very few retailers, if any, bought or sold more than 25% of a product at the national or even at the regional level; second, the amendment made it easier for the government to control horizontal mergers among manufacturers since merger control could now be applied whenever the merging firms had a market share greater than 25% of a market.

A more significant step towards meeting the demands of the manufacturers was achieved by the 1986 ordinance through the adoption of a provision enabling the government to control mergers that are likely to affect competition whenever the combined total sales of the merging firms is larger than seven billion francs (and provided that the pre-merger total sales of at least two of the parties to the merger was larger than two billion francs) even if the combined market share of the firms involved does not meet the 25% requirement. The significance of this change in merger control law cannot be explained by economic reasoning since, as is well known, there is no direct relationship between the absolute size of a firm and its market power. There is thus no *a priori* reason to believe that a merger between two firms might impair competition for the sole reason that their absolute size is large. However, as the total sales of a number of distributors in France exceeds two billion francs, this provision will make it easier for the government to control at least some mergers in the retail sector.

## 1.5.2. Abuse of buying power

The debate on the necessity of finding ways to curb 'abuses of buying power' was triggered in 1983 and 1984 by the fact that some large distributors (acting either individually or through common buying agencies) refused to buy (or rather refused to continue to carry) some brand food products or beverages unless they were granted special additional (discriminatory) discounts by the manufacturers concerned. In each case, the brand products involved accounted for a minute proportion of the total sales of the distributors. It was also obvious that although these distributors held a significant share of the market of the product at the retail level (say between 10% and 20%), they could not be considered to have a dominant position either as buyer or as seller.

Yet, for a manufacturer the sales of the brand product to one of those distributors could represent a major part of its total sales.

The manufacturers claimed they had no choice but to grant the discounts requested by the distributors (or by the common buying agencies) and thus ultimately face bankruptcy.

With this controversy in mind, and because it also wanted to alleviate what it considered a growing imbalance in the economic relationship between manufacturers and large distributors, the Chirac administration introduced a provision in the 1986 ordinance making it illegal for firms (even if they do not have a dominant position on the market) to abuse their economic bargaining power *vis à vis* other firms (suppliers or customers) when those other firms are dependent on them and when the abuse constitutes a restraint of competition. The ordinance states, among other things, that the discontinuation, by a firm, of an established commercial relationship with a dependant firm because the latter refuses unjustified commercial demands could be considered such an abuse.

The introduction of this provision in the ordinance on competition suggests two comments.

First, one could argue that the fact that a firm (manufacturer A) is dependant on another firm (one of its suppliers or one of its distributors: firm B) tells us nothing about the general market power of firm B and the ability of firm B to impair competition. Therefore, there is no reason to think *a priori* that the behavior of firms holding other firms in their dependency is more likely to be anticompetitive than the behavior of firms which do not hold other firms in their dependency. Thus the logic of this provision is very remote from the logic of competition and might even be considered in contradiction with it if it were used to freeze commercial relationships between suppliers and customers.

Second, the concept of dependency between a supplier and a buyer is, to say the least, a difficult one to define. If a supplier has chosen to sell all of his output to one particular distributor and has adapted its products to fit the needs of the customers of this particular distributor when he could have chosen a different strategy (such as selling its products to a variety of distributors) should he be considered as dependent on his distributor in spite of the fact that his lack of strategic flexibility (at least in the short run) is the result of his own doing?

### 1.5.3. Individual practices

The question of whether to do away with the four main prohibitions of 'individual practices' (price discrimination, refusal to deal, resale price maintenance, reselling at a loss) or to keep them on the books was hotly debated in the government and business communities.

Because the government had at long last realized the difficulties and the inadvisibility of enforcing the *per se* prohibition on price discrimination, it was decided, at the end of 1985, to amend the 1945 ordinance on this point. From December 1985 on (loi No. 85-1408 du 30 décembre, 1985), price discrimination was prohibited only if it was likely to impair competition. This change in legislation which gave more freedom to suppliers in their pricing policy was, however, opposed both by small distributors and by some manufacturers of standardized brand products (in particular in the food and beverage industries) because they felt that the relaxing of the rule put them in a weaker position to resist the demands of price discounts by powerful distributors. They argued that, contrary to government belief, what was actually needed was a more rigid enforcement of the *per se* prohibition on price discrimination and stiffer penalties.

This debate continued with renewed intensity during the preparation of the 1986 ordinance on price freedom and competition. The new (Chirac) administration, which felt that *per se* prohibition of price discrimination was basically inconsistent with its aim to promote competition and freedom of pricing gave in neither to the demands of manufacturers worried by the buying power of some distributors, nor to the demands of traditional small-scale distributors. It argued that new means to fight illegitimate demands by large-scale distributors had been written into the ordinance (through the prohibition of abuses of situations of dependency and by the possibility of controlling mergers in the retail sector) and that a return to the *per se* prohibition of price discrimination was therefore unnecessary.

The abandonment of *per se* prohibition of refusal to sell in the 1986 ordinance on price freedom and competition can be considered as a victory for the manufacturers because such a prohibition was considered to be contradictory to the ordinance's main goal of promoting economic freedom.

On should note, though, that unjustified discriminatory practices on the part of suppliers, unjustified discriminatory demands on the part of buyers, and refusal to sell, remain civil offences under French law when they impair competition.

The *per se* prohibition of two other 'individual' practices (resale price maintenance and reselling at a loss) were also maintained in the 1986 ordinance.

The justification offered for the *per se* prohibition of resale price maintenance was that it restraints intra-brand competition at the retail level. One may assume that beyond this rather weak justification, public policy makers still have the misguided impression that allowing this practice would fuel inflation.

Although *per se* prohibition of reselling at a loss is clearly anticompetitive, this measure was favored as an attempt to protect traditional small distributors from the rigors of competition with the larger distributors. Heated discussions on this subject centered around two questions: first, whether distributors should be permitted to resell at a loss when they want to meet the price of a competitor; second, what is meant by reselling at a loss or, in other words, whether year-end quantity discounts could be deducted from invoice prices to determine the actual purchase price below which a good cannot be resold.

On the first point, it should be noted that distributors are permitted to resell at a loss only to meet the 'legal' resale price of a competitor.

On the second point, the ordinance states that the price to be taken into consideration when determining minimum resale price is the invoice price (to which must be added sales tax and transportation cost). This means that the only discounts or rebates which can be taken into consideration are those which the distributor is entitled to at the moment of his purchase.

The provision regarding the *per se* prohibition of refusal to resell at a loss thus reflects a partial victory for small distributors who actively lobbied to increase the minimum resale price of their larger competitors.

## 2. THE ENFORCEMENT OF COMPETITION LAW IN FRANCE

Having reviewed the history of competition law in France, we now turn to the enforcement of competition statutes since 1977.

At the outset it is worth mentioning a few general features of the enforcement process.

First, one should note that the majority of the cases brought to the Commission de la concurrence and the Conseil de la concurrence have been cases involving 'concerted actions, conventions, explicit or tacit understandings or coalitions which are designed for or may have the effect of curbing, restraining or distorting competition'. Those practices are now prohibited by article 7 of the 1986 ordinance and were formerly prohibited by article 50 of the 1945 ordinance.

Most of those cases have been horizontal (agreements between potential competitors), but a significant number have been vertical cases (and in particular cases relating to selective distribution contracts).

In contrast, it is worth mentioning that cases involving abuses of market power by firms holding a dominant position (now falling under article 8 of the 1986 ordinance) have rarely been sent to the Conseil de la concurrence.

Similarly, as was mentioned earlier, only about fifteen merger cases have been brought to the Commission or to the Conseil over a twelve-year period.

Second, one should point out that until 1986 the Commission de la concurrence, and since then the Conseil de la concurrence, have played a major role in interpreting the ordinances prohibiting anticompetitive agreements and abuses of dominant positions. Up until 1977, firms and trade organizations whose practices were found in violation of the law could file an appeal against the decision taken by the minister in charge of economic affairs (after the Commission de la concurrence had issued its opinion on the case) with the Administrative Supreme Court. Since 1986, the Paris Court of Appeals reviews the decisions of the Conseil de la concurrence.

In the pre-1986 system, the Minister in charge of economic affairs generally accepted the findings and the analysis of the Commission (although he felt free to inflict more lenient penalties than those recommended by the Commission) and the Administrative Supreme Court restricted itself either to determining whether the facts of the case had been clearly established or to procedural matters but it never questioned the interpretation of the Commission or the Minister as to what constituted an anticompetitive practice.

One of the few instances in which the Administrative Supreme Court overturned a ministerial decision concerned a case in which a number of perfume manufacturers had simultaneously refused to sell to a discounter. The Commission considered that this parallel behavior implied a tacit agreement among the manufacturers despite their claim that they had not entered into such an agreement and that each one of them had had an independent reason to refuse to sell to the discounter[2]. The Commission based its reasoning for establishing that there had been a tacit agreement among the manufacturers on the fact that the reasons given by each one of them to explain its (independant) refusal to deal with the discounter were either unsatisfactory or contrary to the facts of the case. Thus the Commission considered that the manufacturers had entered a tacit agreement with the purpose and the effect of restraining price competition among their distributors and it suggested that they should be fined. Following the recommendation of the Commission, the Minister imposed a fine on the manufacturers. However the Administrative Supreme Court ruled that the fact that each manufacturer had been unable to satisfactorily explain why it had refused to sell to the discounter was insufficient to establish a meeting of minds and therefore an agreement (explicit or tacit) among them.

Similarly, since 1986, the Paris Court of Appeal has confirmed most of the decisions of the Couseil de la concurrence and has never reversed a decision of the Conseil because of disagreement with its interpretation of the substance of the ordinance.

Third, to determine whether a particular agreement is or is not prohibited by article 7 of the 1986 ordinance, the Conseil de la concurrence generally turns its attention to three issues:

Does the practice violate the condition of independence of decision-making by the firms on the market?
Does the practice diminish the uncertainty which each firm should face regarding the strategy likely to be followed by other firms?
Does the practice in any way impede the ability of other firms to enter the market or to expand on the market?

It can thus be said that the Conseil looks at the process through which prices and quantities are determined rather than at the result

---

[2] Décision No. 84–13/DC du ministre relative à la situation de la concurrence dans le secteur de la distribution sélective des produits de parfumerie, *Rapport de la Commission de la concurrence pour l'année 1984*, p. 63.

achieved. For example, in a 1987 decision[3] concerning bakers, the Conseil found that in one area the professional organization of bakers had urged its members to increase the price of bread by a certain amount in January and July, 1986. It was also established that most bakers had indeed followed the recommendation of their professional organization. The Conseil thus considered that the professional organization had initiated an anticompetitive agreement among its member firms. In so doing it refused to take into consideration a claim by the organization that in other areas (in which there was no evidence of an agreement among bakers) the price of bread was higher (or increased faster) than in the area considered. The reasoning followed by the Conseil was that the actual level or rate of increase of the price of bread elsewhere was irrelevant since it was established that the process through which it was arrived at was not independent decision-making.

Similarly, except in cases involving price discrimination, the Commission and the Conseil have refrained from using the absolute levels of prices or profit rates as evidence establishing the anticompetitive nature of an abusive practice by a firm having a dominant position. The caution exercised by competition authorities in this area can be explained both by their process approach to competition and fear of appearing to re-establish price controls under the guise of competition law.

Fourth, until 1986 article 50 of the 1945 ordinance, and since then articles 7 and 8 of the ordinance of 1986, state that agreements whether explicit or tacit among firms etc . . . and abuses of dominant positions are prohibited only if their object was to impair or restrain competition, or if they had or could have such an effect. Thus the onus of proof rests with the competition authority which must establish for each case why the particular practice examined had an anticompetitive effect, could have had such an effect or had such an objective.

Although the wording of the ordinance seems to preclude a *per se* approach, the question of whether competition authorities in fact apply a *per se* approach or a rule of reason approach, largely rests on the kind of reasoning they use to determine whether or not a particular

---

[3] Décision No. 87-D-33 relative à des pratiques relevées dans le secteur de la boulangerie artisanale des Côtes-du-Nord, *Rapport du conseil de la concurrence pour l'année 1987*, annexe No. 42, p. 6.

practice at hand is illegal. If the reasoning is so general that it could be applied to all cases in which the practice was found, the approach can be described as a *per se* approach; if the reasoning is based on specific considerations pertaining to the market examined or to the firms under scrutiny, then the reasoning can be considered to be a rule of reason.

Whenever the Conseil finds that firms have engaged in a practice that *had* the effect or the objective of impairing competition, it follows a rule of reason approach in that its opinion is based on the specific aspects of the market examined. However, in a number of cases in which the Conseil (or previously the Commission) has found firms guilty of having engaged in practices that *could have had* the effect of impairing competition (and therefore relieves itself with the burden of proof that the practice in fact had that effect), it follows a *per se* approach.

The rule of reason approach can typically be found in the Conseil's decisions on vertical restrictive agreements (such as selective distribution contracts), or on parallel price increases in concentrated industries or on cases involving abuses of dominant positions. In such cases competition authorities try to establish whether the practices examined could have restrained competition on the particular markets examined.

But some other practices, such as exchanges of information among potential competitors prior to submissions for public tender offers, are treated as *per se* offences.

Over the years more and more cases have been decided either on the basis of the fact that the objective of the practices examined was (necessarily) to restrain competition or on the ground that such practices could by their very nature impair competition. Thus the *per se* approach has gained ground in many instances even if the competition authorities have always been reluctant to admit it.

Fifth, firms or professional organizations found to have entered into an anticompetitive agreement or to have abused their dominant position can be exonerated if it is established that their practice was the direct (and inevitable) consequence of a law or, in certain cases, if it contributed to economic progress. However the ordinance places the burden of proof on the firms or organizations themselves.

Competition authorities have generally been reluctant to admit that a competitive practice which restrained competition could contribute to economic progress. Indeed, the Commission de la concurrence chose to take a restrictive interpretation of article 51 and to require firms to show first that the alleged economic progress was the direct consequence of the practice examined and, second, that it could not have been achieved by any other means than by the adoption of the practice.

The Conseil de la concurrence has followed the path opened by the Commission de la concurrence regarding the interpretation of article 51 of the 1945 ordinance by imposing the same standards of proof. It should be noted that the efficiency defense is likely to become even more difficult to use in the future because article 10 of the 1986 ordinance, which has replaced article 51 of the 1945 ordinance, restricts the possibility of an efficiency defense to cases in which the anticompetitive practice 'reserved an equitable share of the resulting profits to the buyers' and did not 'enable the businesses concerned to eliminate competition on a substantial portion of the markets involved'.

One of the rare cases in which the Commission de la concurrence applied article 51 of the 1945 ordinance to an anticompetitive price agreement concerned Interflora[4]. In this case the Commission first considered that an agreement between Interflora and its (independent) affiliated florists throughout France to agree on the price that should be charged on orders concerning standardized flower arrangements restricted competition among the affiliates. The Conseil based its reasoning on the fact that a customer from one city who used the Interflora system to order the delivery of flowers in another town was quoted a unique price in spite of the fact that in that town several affiliates could have executed the order and that they could have competed on prices to fulfill the order if they had not been bound by the agreement. However the Conseil also considered that striking down the price agreement among Interflora and its affiliates would lead to a significant increase in information costs for consumers wishing to place a long-distance flower delivery order since they would then have to establish contact themselves with the Interflora affiliates of the town in which they wanted to send their flowers. This increased cost, would in turn decrease demand for the type of service created and provided by Interflora. In other words the Conseil considered that the restriction to

---

[4] Avis de la Commission de la concurrence relatif à la situation de la concurrence dans le secteur de la transmission florale et décision ministérielle, *Rapport pour l'année 1985*, Annexe No. 18, p. 123.

competition was in fact a necessary condition for permitting this particular type of service.

## 2.1. Horizontal agreements

A wide array of practices such as price-fixing among competitors, price recommendations by trade organizations, market-sharing and collusion on tender offers, coordinated attempts to exclude competitors or importers, exchange of informations among competitors, etc . . . fall in this category.

### 2.1.1. Explicit agreements

Over a ten-year period (1977-1988), explicit price-and market-sharing agreements among competitors represent slightly more than a quarter of all the established violations of French competition law. Collusions in tender offers represent roughly 20% of all cases submitted to competition authorities.

Market-sharing agreements, although not uncommon, do not warrant a particular comment other than the fact that they are treated as *per se* illegalities.

Price agreements are more common and also more varied; the approach of competition authorities in this area deserve further comments.

Under the general heading of explicit price agreements, one should distinguish among explicit agreements among competitors to fix prices actually charged, explicit agreements among competitors to determine tariffs when substantial rebates are common practice, establishment and publication of recommended or suggested tariffs by trade or professional organizations, and the exchange of information on future prices among competitors.

*2.1.1.1.* Explicit agreements among competitors to fix prices, price increases or profit margins are treated as *per se* offences at least since 1985. Reflecting its view on this matter in its 1987 annual report, the Conseil de la concurrence stated: 'its strong attachment to independent decision-making in the price area. Indeed, this independence is a necessary condition for the existence of price competition which—although it is but one form of competition—is nevertheless a determining factor in forcing firms to be as efficient as possible'.

French competition authorities consider that such practices violate the law even if the firms party to the agreement represent only a fraction of the suppliers of the good or the service considered on the market. As a matter of fact, in many cases their decision will not give a precise description of the market identifying all the suppliers (although it will always precisely define the product or the service considered). If a *per se* rule has generally been applied to horizontal price-fixing agreements since 1985, this was not always previously the case.

Between 1981 and 1984, the Commission de la concurrence occasionally exonerated some horizontal price agreements (as well as some other horizontal or vertical practices) from the prohibition of article 50 of the 1945 ordinance because it considered that the actual or potential anticompetitive effect of those agreements was not sufficiently significant[5].

*2.1.1.2.* Explicit agreements among competitors to determine tariffs even when substantial discounts are offered by the competitors, or the establishment and publication of recommended or suggested tariffs by trade or professional organizations, have also generally been treated as *per se* offences. Although competition authorities often explain why they think that such practices are prohibited by article 50 of the 1945 ordinance or by article 7 of the 1986 ordinance, the reasoning used is usually general and not specific to the case at hand.

As far as concerted actions among potential competitors to determine tariffs are concerned, competition authorities consider that such actions violate the competition law irrespective of whether or not the prices actually charged to consumers were equal to those mentioned on the tariff. Thus, in a 1985 case concerning cardboard manufacturers[6], the Commission de la concurrence, after having established that the firms involved had held meetings to determine common increases in their tariffs went on to say 'Although it has not been established that this concertation has had an effect (on prices actually

---

[5] Décision No. 81-10/DC du ministre relative à une entente dans la boulangerie dans le département de l'Oise et avis, *Rapport pour l'année 1981*, annexe No. 11, p. 184: Décision No. 81-06/DC du ministre relative à des pratiques concertées dans le secteur de la réparation automobile et avis, *Rapport pour l'année 1981*, annexe No. 7, p. 140.

[6] Décision No. 85-1/DC relative à des pratiques concertées dans le secteur de la production du carton ondulé et avis, *Rapport pour l'année 1985*, annexe No. 9, p. 60.

charged), it remains that the objective pursued was to encourage an artificial price increase (and that the firms have violated article 50 of the 1945 ordinance)'.

The reasoning followed by the Conseil de la concurrence regarding the publication of recommended prices by trade organizations appears quite clearly in a 1987 decision concerning architects[7]. In its decision, the Council explained that 'besides their purpose, such tariffs may have an anticompetitive effect, . . ., in that they may decrease the incentive (of architects) to compute their costs individually (and to determine their prices). This potentiality is independent of the number of (firms) that actually apply the tariffs'.

Commenting on its views on the subject on a more general level, the Conseil de la concurrence wrote in its 1987 report 'the publication by trade organizations of price lists, price recommendations or indications on prices to be charged, is prohibited, even in the absence of pressure on the professionals involved to follow the recommendations, because by giving an indication to all firms about the level of what is considered by the profession as a 'normal' price or price increase, the trade organizations may give an incentive to some or all of those firms to actually adopt those prices or price increases'.

Finally, one should note that horizontal explicit agreement by retailers to boycott manufacturers so as to protect their profit margins is not infrequent and is considered to be illegal *per se*. One celebrated case of this type concerned the relationship between pharmacists and pharmaceutical firms[8].

The profession of pharmacist is strictly regulated in France. Pharmacists have a monopoly on the sale of prescription drugs and the maximum price-cost margin on those products is determined by the government. Finally, pharmacists must respect a code of ethics which, among other things, prevents them from substituting other drugs for drugs prescribed by doctors and prohibits them from engaging in 'non-confraternal' behavior. Pharmacists have always interpreted their code of ethics as meaning that they should not engage in price competition.

---

[7] Décision No. 87-D-53 du Conseil de la concurrence relative à la situation de la concurrence dans le domaine des architectes et avis. *Rapport pour l'année 1987*, annexe No. 62, p. 80.
[8] Avis relatif à des pratiques concertées de pharmaciens d'officine pour s'opposer à la commercialisation de médicaments d'origine et décision No. 81-07/DC du ministre. *Rapport annuel de la Commission de la concurrence pour l'année 1981*, p. 194.

Therefore they stick to the maximum price-cost margin allowed on prescription drugs. At the end of the '70s, when Clin Midi, a large phameceutical firm, decided to launch a line of generic prescription drugs, French pharmacists became alarmed because they realized that given the comparatively low price of generic prescription drugs they risked a decrease in their revenues. As pharmacists can neither increase the price-cost margin charged for those drugs nor refuse to sell them if they are prescribed, they decided, with the help of their trade organizations, to refuse to carry all the Clin Midi non-prescription drugs products that they previously sold. After six months of this concerted boycott, Clin Midi decided to discontinue the production and sale of generic prescription drugs and pharmacists lifted their boycott. Subsequently, the pharmacists' trade organizations were found guilty of having engaged in anticompetitive practices and were fined by the Minister. However Clin Midi never resumed its production of generic prescription drugs.

### 2.1.2. Tacit agreements and parallel behavior

Since 1977, French competition authorities have repeatedly stated that evidence of parallel behavior among competitors (for example simultaneous increases in prices or simultaneous refusal to sell to a discounter or absolute stability of market shares) is not sufficient to demonstrate the existence of a tacit agreement among them and that it must also be established either that 'a meeting of minds' lies behind and explains the parallel behavior observed or that it 'cannot be explained by other considerations'[9].

In various cases, competition authorities considered that the observed parallelism of behavior was insufficient to establish a tacit agreement prohibited by the ordinance because each of the firms involved had an interest in pursuing the practice independently of whether or not its competitors pursued the same practice.

For example, in 1982 the Commission de la concurrence did not consider that the fact that various watch manufacturers had simultaneously refused to sell to a discounter[10] was sufficient to establish a

---

[9] Rapport annuel de la commission de la concurrence pour l'année 1981. *Rapport annuel du Conseil de la concurrence pour l'année 1987*, p. XIV.

[10] Décision No. 82-7/DC relative à des pratiques anticoncurrentielles relevées dans la distribution de certains produits horlogers et avis. *Rapport pour l'année 1982*, annexe No. 11, p. 159.

meeting of the minds among them. Indeed it considered that the pressure exercized by traditional distributors on each of their suppliers because they did not want to carry products if they had to compete on those products with the discounter, gave each of the manufacturers a sufficiently personal reason to refuse to sell to the discounter, irrespective of the attitudes of the other watch manufacturers.

In other cases competition authorities were satisfied that a given example of parallel behavior was necessarily the result of a tacit agreement as no other explanation could be found. One such case occurred in 1983 when the Commission de la concurrence examined the pricing behavior of five brick manufacturers in the south of France who had simultaneously increased their prices in the same proportion six or seven times over a two-year period[11]. In its opinion, the Commission determined that the structures and the costs of the firms were not identical and that the price of substitutes did not prevent the firms from being profitable. It stated 'when firms belonging to the same industry increase their prices simultaneously and more rapidly than their costs increase (which implies that each one of them could have deferred its price increase or increased its prices less than it did) and when, furthermore, those price increases are on numerous occasions simultaneous and strictly identical (which excludes the possibility that they happened by chance or that they resulted from a competitive process), one must conclude that those price increases result from an anticompetitive agreement prohibited by article 50 of the 1945 ordinance'.

However, the number of cases in which competition authorities concluded that a parallel behavior by several firms implied a tacit agreement is relatively small for two reasons.

First, as was mentioned earlier, the Administrative Supreme Court insisted that proof of a meeting of the minds had to be a direct proof and could not be based only on the fact that the firms' explanation of their parallel behavior was unsatisfactory.

Second, the competition authorities have had difficulties applying the 'meeting of the minds' standard to cases of oligopolistic industries (with homogeneous products and near perfect information) in which

---

[11] Décision No. 83-6/DC du ministre relative à la situation de la concurrence dans la production des tuiles et briques dans la région Midi-Pyrénées et avis. *Rapport pour l'année 1983*, annexe No. 10, p. 191.

parallel behavior is frequently observed. Indeed, in those industries, the parallel behavior of firms can be due to the fact that each one decides independently to follow an oligopoly strategy.

Thus, for example, when the Conseil de la concurrence examined the pricing behavior of the two French cigarette paper manufacturers (used by consumers who roll their own cigarettes), it observed that they had identical prices and had increased those prices at the same periods nine consecutive times over a two-year period[12]. If the firm initiating the increase was not always the same, the other firm always followed within a few days. Each of the firms, however, argued that the reason it had always followed the price increases initiated by its competitor was simply that it could not gain anything from not doing so because the only possible result of a refusal on its part to follow its competitor would have been an immediate cancellation of the price increase by the other firm. Thus, they argued they had not 'tacitly agreed' to cancel price competition but had separately recognized their interdependency in this area and had competed through other means, most notably through advertising campaigns aimed at the distributors. Because the evidence showed that distributors and consumers were not very sensitive to the absolute level of price but were very sensitive to price differences between the two manufacturers and because it also showed that market shares were not constant over time in spite of the identity in prices, the Conseil de la concurrence considered that it was not established that they had engaged in a prohibited practice.

### 2.1.3. Exchanges of information

Whereas exchange of informations on prices to be charged are in fact treated as *per se* illegal, the exchanges of information on variables other than prices (such as costs of raw materials, wages etc . . .) are examined through a rule of reason.

### 2.2. Vertical restrictive agreements

For political reasons that have been explained previously, resale price maintenance is prohibited *per se* independently and is handled by the

---

[12] Décision No. 88-D-02 du conseil de la concurrence relative à la concurrence dans le secteur de l'approvisionnement des débits de tabac en fourniture accessoires. *Rapport du Conseil de la concurrence pour l'année 1988*, annexe No. 9, p. 21.

regular courts rather than by the competition authorities.

Other vertical restrictive agreements such as exclusive dealerships or selective-distribution contracts can fall under article 50 or 7 of the ordinances which prohibit all 'conventions' (i.e. contracts) which have the objective or can have the effect of impairing competition.

However, whereas horizontal agreements are mostly treated by competition authorities as *per se* offences, a rule of reason is used to assess the impact of selective or exclusive-distribution contracts on competition.

As far as selective distribution systems are concerned, the analysis of French competition authorities on this subject has always been close to that of the European Court of Justice as expressed in its decision Metro II[13].

Selective distribution contracts have contradictory effects on competition. First, they reduce competition among the distributors of the same product (intra-brand competition) and thus give each of them an incentive to push sales of the product considered (for example through the provision of some specific services that they would not provide if their expected profit margin was lower).

Second, the existence of such restrictive agreements may increase or reflect the intensity of inter-brand competition in various ways.

Third, the efficiency effect of selective-distribution contracts (or other vertical restrictions of trade) depends on how much marginal and infra-marginal consumers value the services provided by the distributors.

The consequences of restrictive-distribution systems on competition and on consumer welfare are thus complex and raise several questions. Do they provide consumers with a broader range of products available? Does the increase in services at the retail level increase consumer welfare? Does the increase in prices stemming from the lessening of competition at the retail level merely equal the cost of the added services to the consumer or does it enable distributors to enjoy monopoly profits?

Competition authorities have not tackled all the afore-mentioned issues when dealing with selective-distribution systems. They have

---

[13] See Metro SB-Grossmaerkete II, European Court of Justice, 75/84, October 11, 1986.

# Now available on continuation order

# FUNDAMENTALS OF PURE AND APPLIED ECONOMICS

*Fundamentals of Pure and Applied Economics* is an international series that will appeal to economists in academia, government and business. New findings by leading experts are published rapidly and concisely at a level accessible to economists outside a given specialty. The series is divided by discipline into sections, each with its own editor, and publishes volumes as they are received. Individual volumes will later be compiled by section, revised, and published, for easy reference, as the *Encyclopedia of Economics*. The sections and editors are listed inside.

# ORDER FORM

**TO:** Harwood Academic Publishers
Marketing Department
P.O. Box 786 Cooper Station
New York, NY 10276
USA

**OR:** Harwood Academic Publishers
Marketing Department
P.O. Box 197
London WC2E 9PX
UK

[ ] Please send me further details of the other volumes in the series.

[ ] Please enter my continuation order to the Fundamentals of Pure and Applied Economics series, commencing volume _____ (ISSN: 0191-1708)

**PAYMENT METHOD:**

[ ] Charge my credit card: [ ] American Express [ ] Visa [ ] Master Card

Account No. _____ Expiry date _____
Signature _____
[ ] Bill my organization, P.O. No. _____
(We cannot bill your organization without a P.O. No.)
[ ] Bill me
Name _____
Affiliation _____
Address _____
Zip/Postal Code _____ Country _____

**Harwood Academic Publishers**
chur · london · paris · new york · melbourne

mainly directed their efforts at establishing rules that would prevent manufacturers from limiting the ability of distributors to determine their pricing policy independently or from rejecting *a priori* low-margin distributors in their networks. Thus, contrary to what economic theory suggests, vertical restraints of trade in general, and selective-distribution systems in particular, have been analyzed primarily through their effects on competition at the distributor's level.

Using this approach, competition authorities ruled that selective-distribution systems fell within the prohibition of anticompetitive agreements in the perfume sector and in the cosmetics sector[14] whereas they had not impaired competition in the tennis-raquet[15] or in the wind-sail sectors[16] (because in the latter sectors only a small number of manufacturers representing a small share of the market distributed their product through selective-distribution systems).

In cases in which a selective-distribution agreement could impair competition, both the Commission de la concurrence and the Conseil de la concurrence have imposed similar constraints: the manufacturer must not impose a resale price; the requirements set by the manufacturer for accepting a distributor must be explicit, objective and verifiable; the manufacturer must not exclude *a priori* any type of retailers which could technically meet the requirements set by the manufacturer; the technical requirements defined by the manufacturers must not be used in a discriminatory way to exclude low-margin distributors.

The case relating to the distribution of cosmetics by pharmacists provides us with a good example of the reasoning of the Conseil de la concurrence.

A number of pharmaceutical firms distribute their most successful and well-known cosmetic products in pharmacies. Typically, their distribution contracts provide that those products are distributed only by pharmacists who, as we saw earlier, do not compete among each other on prices. For a number of years large distributors have tried

---

[14] Décision du conseil de la concurrence relative à la situation de la concurrence dans la distribution en pharmacie de certains produits cosmétiques et d'hygiène corporelle. *Rapport pour l'année 1987*, Annexe No. 24, p. 43.

[15] Avis de la Commission de la concurrence sur la situation de la concurrence dans la distribution des raquettes de tennis et décision ministérielle. *Rapport pour l'année 1984*, Annexe No. 13, p. 112.

[16] Avis de la Commission de la concurrence dans la distribution des planches à voile et décision ministérielle. *Rapport pour l'année 1984*, Annexe No. 14, p. 114.

to gain access to these products and have denounced the selective-distribution systems used by manufacturers as anticompetitive, on the grounds that they unnecessarily suppressed intra-brand competition at the retail level. Although it is likely that the manufacturers refused to sell to large distributors because they knew that as soon as they accepted an order from one of them, pharmacists would refuse to carry their products, there was no evidence of a threat of a concerted boycott analogous to the one that had been found in the generic product case. Thus when the manufacturers' distribution contracts came under the scrutiny of the Conseil de la concurrence, the manufacturers' defense was based on the fact that cosmetic products were dangerous products requiring counseling that only licensed pharmacists were capable of giving. Additionally, the manufacturers claimed that inter-brand competition was strong and that their distribution arrangements did not significantly restrain competition. Finally, the manufacturers also pointed out that the cosmetics they sold through pharmacists were competing with cosmetics from other manufacturers who had chosen not to sell through pharmacists but through large retail stores. The distributors, on their part, claimed that they were willing to abide by any condition imposed by the manufacturer (such as hiring competent salesmen with a degree in pharmacy), that the cosmetics they could sell did not really compete with the cosmetics sold through pharmacies since brand-names differentiated them and were a major determinant of demand.

The Conseil de la concurrence came to the conclusion that the cosmetics sold by manufacturers who distributed only through pharmacists were not close substitutes of other cosmetics distributed through large-scale retailers as evidenced by the fact that there was a persistent wide difference in prices between the two categories of cosmetics. Having thus determined the boundaries of the specific market examined (i.e. cosmetics sold in pharmacies), the Conseil de la concurrence then considered that the clauses of the distribution contracts which reserved the distribution of those cosmetics to pharmacists violated article 7 of the 1986 ordinance on prices for two related reasons. On the one hand, such clauses completely eliminated intra-brand competition since pharmacists did not compete on prices; on the other hand, they prevented distributors who were not pharmacists from carrying such goods even if those distributors were technically able to provide the kind of services (such as counseling) that the manufacturers desired. In its decision, the Conseil ordered the manufacturers to spell out the technical, objective and verifiable requirements to be accepted in their distribution networks and to sell to any distributor (whether or not he was a pharmacist) who met these requirements.

### 2.3. Abuses of dominant position

It is first necessary to recall the texts of article 50 of the ordinance of 1945 and article 10 of the ordinance of 1986 on this subject.

The last paragraph of article 50 of the ordinance of 1945 prohibited 'the practices of a firm or a group of firms having, on the domestic market or on a substantial part of the domestic market, a dominant position characterized by a monopoly situation or by an obvious economic power, when those practices are designed for or may have the effect of curbing, restraining or distorting the regular working of a market'.

Article 10 of the 1986 ordinance prohibits 'the abusive exploitation by a firm or a group of firms of its dominant position on the domestic market or a substantial part of the domestic market when they are designed for or may have the effect of curbing, restraining or distorting competition'[17].

Because these ordinances do not refer explicitly to monopolization or to attempts to monopolize, French competition authorities have had more leeway to censor anticompetitive practices of powerful firms than competition authorities in some other countries.

Four issues are of importance in relationship with those prohibitions:

1. How do competition authorities define a 'domestic market' or a 'substantial part' of such a market?
2. What criteria are used to define the existence of a dominant position?

---

[17] As no final decision on cases of 'abuses of dependency' have been handed out yet by the Conseil de la concurrence, the following commentaries concern only abuses of market power by firms holding a dominant position.

3. What kind of practices are found to be in violation of the ordinances?
4. What kind of remedies can be applied in cases of abuses of power by firms holding a dominant position?

### 2.3.1. Definition of relevant markets

Competition authorities use the traditional criteria of economic analysis to define a market, although hard data on cross-price elasticities between various goods or services are rarely available. They thus take into consideration factors such as the technical possibility to satisfy similar demands by various means[18], similarities in prices to consumers, identity in the determinants of demand over time[19], differentiation of products or of distribution channels[20], brand differentiation[21], availability of the products or services offered by different suppliers for a group of consumers, etc . . .

French competition authorities have generally tended to define markets narrowly. This is particularly obvious for cases in which some brand products of a particular kind are sold through certain distribution channels (such as traditional distributors) whereas less-notorious products of the same kind are sold through other distribution channels (such as discounters). In these cases, competition authorities have tended to consider that the combination of brand differentiation and differentiation of distribution channels reduces competition between products to such an extent that they can be considered to be on different markets. Thus, for example, the Commission de la concurrence concluded in one case that manufacturers selling pickles and mustard under their brand-names through neighbourhood grocery stores were not on the same market as manufacturers selling mustards and pickles through discount stores under the distributor's brand-name.

---

[18] Avis de la Commission de la concurrence à des pratiques anticoncurrentielles dans le secteur de l'assurance construction et décision ministérielle. *Rapport pour l'année 1980*, annexe No. 6, p. 146.

[19] Avis de la Commission de la concurrence relatif à la prise de contrôle de Locatel par Thorn Electrical Industries. *Rapport pour l'année 1980*, annexe G, p. 89.

[20] Décision 87-D-15 relative à la situation de la concurrence dans la distribution en pharmacie de certains produits cosmétiques et d'hygiène corporelle. *Rapport pour l'année 1987*, annexe No. 24, p. 43.

[21] Avis de la Commission de la concurrence relatif à l'abeorption de la S.E.G.M.A. par le groupe Générale Occidentale. *Rapport pour l'année 1980*, annexe H, p. 90.

For a number of years the Commission de la concurrence hesitated as to the proper definition of markets in geographical terms when it came to establishing whether or not a firm had a dominant position. The difficulty arose from the fact that in certain cases it was obvious that consumers did not or could not substitute a particular kind of good or service provided by suppliers in a geographical area for goods or services of the same kind provided by suppliers established elsewhere. However, as was mentioned earlier, the 1945 price ordinance referred to anticompetitive practices of firms holding a dominant position on 'a domestic market or a substantial part of a domestic market'. The legal question was then whether or not a market that happened to be restricted geographically could be considered to be a 'domestic' market or whether the law only applied to nation-wide markets.

The hesitation of the Commission is clearly visible in its 1979 report which indicates: 'The law states that it is possible to control the practices of a firm holding a dominant position on a substantial part of a domestic market. If the law is interpreted in a narrow sense, it could seriously hamper the control of practices of firms holding a dominant position at the local level, which is a common situation particularly in trade and services. However the Commission has considered in one case that if by nature, or because of legal constraints on the mobility of consumers across suppliers, a market is limited geographically to an area, the law can be applied to a firm holding a dominant position on the local market only if it also holds other local dominant positions on a substantial part of the domestic territory'. To arrive at this conclusion, however, the Commission used a rather confusing approach, first establishing the existence of separate 'local' markets on which the firm examined had a dominant position and then aggregating those local markets into a 'domestic' market on which the firm also had a dominant position because of the large number of local dominant positions it held.

Reasonings of this nature were applied to two cases, one dealing with water distribution systems[22] and the other dealing with undertakers' services[23]. In France, each municipality can choose to grant a

---

[22] Avis de la Commission de la concurrence relatif à des pratiques constatées dans le secteur de la distribution de l'eau et décision ministérielle. *Rapport de la Commission de la concurrence pour l'année 1984*, annexe No. 4, p. 108.

[23] Avis de la Commission de la concurrence concernant le secteur des pompes funèbres et décision ministérielle. *Rapport de la Commission de la concurrence pour l'année 1979*, annexe No. 12, p. 60.

concession for the exploitation of one or the other (or both) services to a private firm. In such cases, consumers are faced with a monopoly since either for technical reasons (in the case of water) or for legal reasons (in the case of undertakers' services) they cannot contract with a firm established in a town other than that in which they live. The Commission then held that each local market was separate from the others because consumers could not substitute services offered elsewhere for those offered locally. However, it did not stop there for its assessment of dominant positions. In each of the cases, it considered that since the same firm had been chosen by a number of municipalities that firm held a dominant position on the 'domestic market' of those services, thereby contradicting its earlier analysis.

However, over the years French competition authorities came to accept more readily the idea that a market separate from other markets could be considered to be a 'domestic' market even if it happened to be a local one. Thus, in 1986, in a case concerning the local and regional press[24], the Commission de la concurrence considered that the different local editions of the same daily newspaper were not substitutes because readers of one town were not interested in the local news of another town (and because each edition was sold only in the relevant town). It thus held that the newspaper involved operated simultaneously on different markets and examined whether or not it had a dominant position on each of those markets.

### 2.3.2. *Criteria used to establish the existence of a dominant position*

The criteria used by competition authorities to define the existence of a dominant position on a relevant market by a single firm do not warrant a long commentary. The importance of the market share is examined both in absolute terms and relatively to the market share of the competitors of the firm considered. However, a large market share is considered in itself insufficient to establish a dominant position.

In addition to market share, competition authorities also take into consideration factors which may affect the possibility for competitors to develop their market shares or for potential entrants to actually

---

[24] Avis de la Commission de la concurrence sur les conditions d'application du dernier alinéa de l'article 50 de l'ordonnance No. 45-1483 du 30 juin, 1945 à des pratiques d'abaissement sélectif de prix dans le secteur de la presse quotidienne d'information locale. *Rapport de la Commission de la concurrence pour l'année 1986*, p. 12.

enter the market considered: upward or downward vertical integration of the firm under investigation, superior management, technical superiority, or product and image differentiation. Thus, for example, the Commission de la concurrence considered that Interflora had a dominant position on the long-distance flower-selling market[25] not only because it had a large market share (eight times larger than the share of its main competitor) but also because it had a strong image and because it had contracted on an exclusive basis with the best-located flower shops in France (before the appearance of its competitor) making it extremely difficult for this competitor to develop its market share even if it managed to increase its own network of affiliated flower shops.

Beyond these criteria, to assess the existence of a dominant position, French competition authorities also take into consideration whether or not the firm examined belongs to a large financial group or holds monopoly power on unrelated markets. Competition authorities assume implicitly that firms with significant market shares and large financial means are perceived by potential entrants or smaller competitors as likely to use predatory pricing to prevent the emergence of competition. Thus barriers to entry in an industry are assumed to be more important (and therefore the position of an already established firm with a large market share more dominant), *ceteris paribus*, when the firm under consideration is 'financially' powerful. The wisdom of such an approach will be discussed below when we consider how competition authorities define 'abuses' of firms holding a dominant position.

Before turning to this question, a last comment should be made about the definition of dominant firms. The 1945 and 1977 ordinances prohibit anticompetitive behavior by firms 'or group of firms' holding a dominant position on a market. There has been a fair amount of controversy on the definition of what constitutes a 'group' of firms. The main question raised was whether or not a tight oligopoly could be considered to be a 'group of firms' holding a dominant position. The Commission de la concurrence did not hesitate to give a positive answer

---

[25] Avis de la Commission de la concurrence relatif à la situation de la concurrence dans le secteur de la transmission florale et décision ministérielle. *Rapport de la Commission de la concurrence pour l'année 1985*, annexe No. 18, p. 123.

to that question in 1979[26]. This rather loose interpretation of the law enabled the Commission to declare illegal what it considered to be parallel anticompetitive behavior in oligopolies even when there was no evidence of an explicit or tacit agreement between the firms involved. However, in recent years, competition authorities have considered that different firms held a joint dominant position only in two cases: when the firms involved belonged to the same financial group or when they had entered an anticompetitive agreement[27].

### 2.3.3. Anticompetitive practices of firms holding a dominant position

A large number of cases in which a firm (or a group of firms) holding a dominant position was found to have engaged in anticompetitive practice involved alleged attempts to drive a competitor off the market or to prevent entry through predatory pricing behavior.

Empirical evidence seems to suggest that well-entrenched, financially powerful firms are occasionally willing to follow such a strategy when confronted with new competitors.

While some may argue that the adoption of a low-price strategy by a dominant firm confronted with a new competitor results in an improvement of the price/quality ratio for consumers, and therefore increases consumer surplus, it would seem that the positive effect of such a strategy can be short-lived if the entrant is driven off the market. What is more, if successful vis-à-vis one particular entrant, this strategy can also have the long-run effect of discouraging any other potential entrant from trying to enter the market. Thus the short-run advantages to competition must be compared to the long-run disadvantages.

An example of such a case was examined by the Commission de la concurrence in a general opinion concerning the daily regional and

---

[26] Avis de la Commission de la concurrence concernant la diffusion de films cinématographiques et décision ministérielle. *Rapport pour l'année 1979*, annexe No. 14, p. 176.

[27] See, for example, Avis de la Commission de la concurrence relatif à la situation de la concurrence sur le marché des treilis soudés et décision ministérielle. *Rapport pour l'année 1985*, annexe No. 13, p. 90.

local press[28]. In this sector a small number of powerful regional newspapers publish several local editions, which each have dominant position or a monopoly in an area or in a given town. Because these large regional newspapers do not usually invade each others territory, the main source of potential competition in a given town comes from small local newspapers with limited ability to incur losses for a long period of time without going bankrupt. Thus, when such competitors appear or when small, already-established newspapers try to increase their circulation, the dominant firm can be tempted to selectively increase the number of pages of the local edition concerned, to decrease its price for consumers, and to reduce advertising rates, thereby pushing the entrant towards bankruptcy.

The Commission de la concurrence qualified its censorship of such practices by stating that if price discrimination among local editions of the same newspaper was not objectionable in itself, it could become a prohibited practice for a firm holding a dominant position when the objective pursued by the firm was not to meet competition but to prevent entry or to drive a competitor off the market. The Commission added that only a case-by-case examination of the facts could enable it to decide whether a price decrease by a dominant firm faced with competition improved competition or hampered it.

The Commission de la concurrence also examined a case of predatory behavior in a related sector ('shoppers')[29]. In that case the dominant firm explicitly tried to drive off the market two newly established newspapers in two different areas by using discriminatory low advertising rates (well below average costs). The small entrants did go bankrupt whereupon the dominant firm increased its advertising rates by 400% in one area and by 1500% in the other. This behavior was found to be in violation of the prohibition of abuses of dominant position.

---

[28] Avis de la Commission de la concurrence sur les conditions d'application du dernier alinéa de l'article 50 de l'ordonnance No. 45-1483 du 30 juin, 1945, à des pratiques d'abaissement sélectif de prix dans le secteur de la presse quotidienne d'information locale. *Rapport de la Commission de la concurrence pour l'année 1986*, annexe No. 1, p. 12.

[29] Avis de la Commission de la concurrence dans le secteur de la presse gratuite en région Provence Alpes Cote d'Azur et décision ministérielle. *Rapport de la Commission de la concurrence pour l'année 1986*, annexe No. 7, p. 51.

In another case the Commission de la concurrence found a group of dominant firms guilty of abuse of market power for having used predatory pricing behavior to force smaller competitors to enter a price agreement with them[30]. In that case, which involved the four largest manufacturers of steel rods used for strengthening concrete, the dominant firms happened to be subsidiaries of the two nationalized steel-makers. Having entered a price agreement with Italian importers, they quickly realized that the Italian importers were 'cheating' by offering discounts. The dominant firms then drove the price of steel rods down to a level below the price of the steel used to manufacture them for a period of six months. During this entire period the Italian importers stayed out of the French market. Having shown their strength (which was largely based on the fact that they did not care about losses because of their status as public firms) the domestic firms allowed the Italians back on the French market in the context of a new price agreement.

### 2.3.4. Remedies in cases of abuses of dominant position

In most of the cases involving abuses of dominant position, the competition authorities issued an injunction prohibiting the firm or the group of firms involved from continuing the abusive practice and inflicted a fine.

The ordinance of 1986 (as the ordinance of 1945 did previously) stated that structural remedy (such as the dismemberment of the the firm involved) can be used only when the dominant position of the firm was acquired through merger and when there is proof that the abuse of dominant position is an inevitable consequence of the merger that gave the firm its dominant position. In such a case, competition authorities cannot act directly but can ask the Minister of Finance to undo the merger that led to the acquisition of a dominant position. Because French competition authorities never came across a case where they felt that the abusive practice was an inevitable consequence of the dominant position of the firm involved (and because in many cases the dominant position itself did not result from a merger) they never asked for a structural remedy.

---

[30] Avis de la Commission de la concurrence relatif à la situation de la concurrence sur le marché des treillis soudés et décision ministérielle. *Rapport de la Commission de la concurrence pour l'année 1985*, annexe No. 13, p. 90.

## 2.4. Merger control

Since 1977 merger control has not been an important tool in the enforcement of competition law in France. On average, hardly one merger a year has been sent for review to the Commission or to the Conseil de la concurrence by the Minister in charge of economic affairs, who has sole authority to initiate the control procedure.

If the French government has progressively abandoned the aggressive pro-merger industrial policy it favored in the '60s, there is still a lingering feeling among public officials that large firms are better able to withstand international competition than smaller firms and that therefore, in general, one should not interfere with the attempts of French firms to increase their size through mergers.

Most of the mergers which have been sent for review to the Conseil de la concurrence over the last few years involved a foreign firm acquiring either a French firm or the French subsidiary of another foreign firm. This led to claims that merger control was in fact used by the Minister in charge of economic affairs to further protectionist aims rather than to further competition. Thus, for example, the acquisition of Spontex by Minnesota Mining and Manufacturing was opposed by the minister in 1989 (in spite of the fact that the Conseil had considered that its disadvantages from the point of view of competition were outweighed by its advantages from the point of view of economic progress).

The previous mergers for which the Minister in charge of economic affairs had initiated the control procedure involved the acquisition of the French assets of Rowntree Mackintosh by Nestle in 1989 (the Conseil did not find the merger anticompetitive and the minister did not oppose it), the acquisition of Saint Louis (a large sugar manufacturer) by Ferruzi (an Italian agro-food conglomerate) in 1988 and the creation of a joint venture between Colgate Palmolive and Henkel-France (French subsidiary of a German firm), also in 1988. In the latter two cases, the firms decided to abandon their planned mergers (although the Conseil de la concurrence had considered that the Colgate–Henkel project to create a joint subsidiary did not qualify as an anticompetitive merger).

No mergers were examined by the Commission de la concurrence in 1985 and in 1986; but it had examined two mergers in 1984: the acquisition of Duolite International by Rhom et Hass France in the chemical sector (the Commission found that the advantages of the acquisition from the point of view of economic progress outweighed its disadvantages from the point of view of competition and recommended that it be allowed; the minister did not oppose the merger) and the acquisition of Ashland Chemical France by Cabot Corporation in the carbon black industry. In the latter case the Commission de la concurrence considered that the disadvantages of the merger outweighed its advantages and recommended that the merger be prohibited. The Minister opposed the merger but his decision was overturned by the Administrative Supreme Court on procedural grounds.

Whatever one may think of the goals pursued by the Minister of economic affairs in deciding which mergers to control, the competition authorities have tended to stick to strictly technical considerations in their opinions, by seeking answers to four different questions.

The first question is whether or not the merger can be controlled: that is whether the share of the merging firms exceeds 25% (previously 40%) on any market or whether the absolute size of the combined firm exceeds the prescribed level. Thus for example, in 1979, the Commission de la concurrence concluded that a merger between Thorn Electrical Industries (a British manufacturer of TV sets) and Locatel (a TV rental firm) could not be controlled by the Minister because they were operating on the same market (and not on different markets as the government claimed) and did not meet the 40% market-share criterion[31].

The second question is whether the merger could have the effect of restraining competition. To answer this question, the competition authorities have taken into consideration concentration on the supply and the demand side, barriers to entry, etc . . . to assess whether or not it is likely that the merging firms will be able to increase prices as a result of the merger. Thus, for example, in its opinion concerning the proposed merger between 3M and Spontex (two manufacturers of sponges for domestic use having a combined market share of 75%) in 1987, the Conseil de la concurrence based its opinion that the merger was unlikely to reduce competition on three main facts: first, there were no significant barriers to entry from the point of view of

---

[31] Avis relatif à la prise de contrôle de Locatel par Thorn Electrical Industries Ltd. *Rapport de la Commission de la concurrence pour l'année 1980*, annexe G, p. 89.

technology, scale economies or capital costs; second, it appeared that a significant portion of the sponges were distributed through large distributors who wielded significant bargaining power; third, the demand for sponges was bound to decrease with the appearance on the market of new products, paper-derived, manufactured by foreign firms[32].

When it finds that a merger could decrease competition, the Conseil de la concurrence must then ask itself whether or not the merger is likely to contribute to economic progress. Thus it examines whether or not the merger will lead to a decrease in unit costs of production or distribution and to capacity expansion. Looking at the merger from a more dynamic point of view, it also assesses whether or not the merger is likely to result in an increase in the rate of innovation or will contribute to the diffusion of technological advances. Thus, for example, when the Commission de la concurrence examined the proposed merger between Duolite International S. A. and Rhom et Haas (two manufacturers of synthetic resins used in water treatment) in 1984, it considered that the merger contributed to economic progress both because the reorganization of production would enable the firms to decrease their unit costs by 5-7% and because technological innovations developed by one of the firms would be transferred to the other[33].

Finally, the last question competition authorities must answer, and by no means the easiest one, when they have found that an anticompetitive merger contributes to economic progress is whether or not the advantages of the merger outweigh its disadvantages.

## 3. CONCLUSION

Recent changes in French competition law must be regarded as a significant step toward promoting competition on French markets to the extent that they abolish price controls, refer the enforcement of

---

[32] Avis No. 89-A-05 du Conseil de la concurrence relatif au projet de concentration entre les sociétés Spontex et 3 M France dans le secteur des outils d'entretien ménager. *Rapport du Conseil de la concurrence pour l'année 1989*, annexe No 72.

[33] Avis relatif au projet de prise de contrôle de la société Duolite International S. A. par le société Rohm and Haas France S. A., *Rapport annuel du Conseil de la concurrence pour l'année 1984*, annexe No. 4, p. 42.

competition law to the courts or to an independent quasi-judicial body, and eliminate a number of constraints on the market strategies of firms which previously hampered competition rather than enabling it to run its course.

The importance of these changes is underlined by the fact that they cannot be considered to result from a short-run change in the political persuasion of the administration. Rather, these changes are the natural continuation of past changes in our legislation and thus reflect a long-run trend towards recognition of the limits of government intervention and of the importance of competition in the market place to insure an efficient allocation of resources.

The importance of the shift towards a stricter enforcement of the ban on anticompetitive practices can be illustrated by the striking increase in the total amount of fines inflicted on firms violating French competition law. These fines went from 4.3 million francs in 1987 to 358 million francs in 1989.

France now considers the fight against anticompetitive agreements and abuses of monopoly power to be a particularly important feature of its economic policy, although it has largely neglected the control of structures. At the same time analysis of the way competition authorities have handled cases reveals that an initial tendency to use competition law to regulate markets has given way to an approach which is both more severe (as is clear from the increased frequency with which violators are fined) and less interventionist. This shift has parallelled the increased sophistication of competition authorities in applying economic theory to antitrust cases.

Because there is now a political consensus as to the usefulness of competition and economic freedom, it is unlikely that this trend will be reversed in the near future. What is more, the acceleration of the construction of Europe in 1993 will mean that the efforts of the Conseil de la concurrence to maintain or restore competition will be supplemented by the spontaneous emergence of new competitive forces, particularly in sectors that have, up to now, remained regulated at the domestic level and been sheltered from the rigors of competition (such as banking, insurance, telecommunication or the airline industries).

# Competition Policy in West Germany: A Comparison with the Antitrust Policy of the United States

ERHARD KANTZENBACH
*Institut für wirtschaftsforschung, Hamburg*

## 1. HISTORICAL EXPERIENCES

In 1890, when the US Congress passed the Sherman Act, Continental Europe as well as the United States was experiencing a fast process of industrialisation and economic growth. This process led to increasing competition in many industries and, as a response to that, to growing efforts to restrain this competition by cartel agreements, by monopolization, and mergers. But contrary to the United States, in Germany no legal action was taken against restraints of competition. In particular the German Supreme Court decided in 1890 that the freedom of contracting applied also to cartel agreements[1].

The result of this judicial decision was a fast growing net of cartel agreements, which after a few years spread over large parts of the economy. In 1905 an official investigation of that matter came to the conclusion that there were already 385 cartel agreements in existence with 12,000 member plants. This situation did not change very much during the period of German empire before the First World War and during the democratic republic and the Fascist dictatorship between the two World Wars, with one exception, that the influence of state planning on the economy was growing fast.

So there was no systematical competition policy until 1957, when in the Federal Republic of Germany the Law Against Restraints of Competition was passed by the federal legislature[2]. And even then, at the

---
[1] Möschel (1983), p. 18.
[2] Möschel (1983), p. 20.

beginning, the policy was rather incomplete, since the law prohibited cartel agreements and abusive practices by dominant firms but did not restrict mergers at all.

In 1973 finally, our law was amended and a merger control was introduced. So our experience with this, in my opinion most important instrument of modern competition policy, has been no longer than fifteen years. This of course is much less than the hundred years of experience with the Sherman Act and even forty years with the Celler-Kefauver Amendment of the Clayton Act.

## 2. BASIC PHILOSOPHIES

There is a fundamental difference in the basic philosophy of the American antitrust law on the one hand and the British and French competition law on the other hand.

The American law is based on the general assumption that free competition is in the public interest. Therefore the general purpose of the antitrust laws is to inhibit any kind of restraints of competition. This may be done by *per se* prohibition of certain market behavior. With respect to other kinds of behavior the rule of reason may be applied. But even in the latter cases there is no room for balancing the goal of free competition against possibly conflicting goals such as the promotion of exports or employment, for example. The rule of reason is only applied to the question whether or not a certain market behavior has under given circumstances restrictive effects.

The basic assumptions on which the British and French competition laws are based are fundamentally different. According to them competition is basically neutral. It may be in the public interest or it may be against it, due to the special circumstances. Therefore cartels, mergers or other types of restraints of competition are judged with respect to their expected economic effect, whether they are expected to promote employment, innovations or international competitiveness or to have negative effects on supplies for the consumer.

With respect to this basic question, the German law follows the American pattern[3]. This is already expressed in its name: 'And against

---

[3] For a comparison in detail of the German and the American law see: Schmidt, Ingo (1973, 1983), Edwards (1978), Mestmäcker and Richtes (1983).

Restraints of Competition'. Under this law, the Federal Cartel Office has the authority only to inhibit different types of restraints of competition, which are enumerated in the law. It does not have the power to balance these against competing economic or social goals. Only two exemptions to this principle exist: in Section 8 and Section 24. Under certain circumstances it is possible to permit, because of public interest, cartels and mergers, which are forbidden under other provisions of the law.

But the power to grant this exceptional permission is given to the Minister of Economic Affairs (Section 8 and 24 III) as the political institution, not to the Federal Cartel Office with its semijudicial functions.

After these more general remarks I should like to discuss some of the instruments of the German competition policy in more detail. In my opinion there are three instruments which were most important for the economic results of this policy, and these are:

first, the prohibition of cartel agreements according to Section 1 of our law,
secondly, the control of the market behavior of market-dominating enterprises according to Section 22, and
thirdly, the merger control according to Section 24 of our Law Against Restraints of Competion.

These three instruments may be compared to Sections 1 and 2 of the Sherman Act and to Section 7 of the Clayton Act, respectively.

## 3. PROHIBITION OF CARTEL AGREEMENTS

Section 1 of the German Law against Restraints declares all agreements by enterprises as void which are likely to restrain competition. Under Section 38 of the law, violators may be fined DM 100,0000 or up to three times the additional profits obtained as a result of the restraint of competition. Because of a very restrictive interpretation by the German Supreme Court this section was not very effective until 1973. The Court stated that an agreement within the meaning of Section 1 could only be found if,

first, the agreement had the property of a private law contract, and secondly, if the restraint of competition was an explicit element of that contract.

This interpretation led to the result that neither concerted actions nor contracts like open-price systems were prohibited.

But these deficiencies have been eliminated now, partly by amending the law, partly by changing the jurisdiction of the courts. Today the prohibition of cartel agreements under Sections 1 and 25 of the German law, Section 85 of the Rome treaty of the European Community, and Section 1 of the US Sherman Act are very similiar. In all three statutes any kind of agreement between competitors is forbidden whose intended effect is a restraint of competition.

But contrary to the American law there are some exceptions from the general prohibition to be found in the German law. According to Sections 2 to 8 two groups of cartel agreements may be authorized by the Federal Cartel Office,

those which are supposed to improve the competitive process, and
those which are supposed to improve the efficiency of the firms concerned.

The first group includes agreements on general terms of business, delivery and payment (Section 2), agreements concerning rebates (Section 3), and those 'necessary to bring about a planned adjustment of productive capacity to demand' (Section 4). These cartels are supposed to restrain those forms of competition which do more harm than good to the economy, or, as we may say, those forms which are not workable in the sense that they do not yield to good economic performance.

The second group of cartel agreements which may be authorized by the Cartel Office are the so-called rationalisation cartels (Section 5). Their exemption from the general prohibition is based on the assumption that restraints of competition may lead to more effective use of resources. The authorization of these cartels therefore requires from the Cartel Office a comparison of the probable effect of the cartelisation with the probable effects of competition with respect to specific technical, economic or organisational functions.

In 1965 and in 1973, the law was amended and the exemptions from the cartel prohibition were extended to specialisation cartels (Section 5a) and to rationalisation cartels of small and medium-size firms

(Section 5b), provided that substantial competition continues to exist in the markets.

Especially, authors of the neo-liberal school criticized the law for its many exemptions of the prohibition principle. Some of them argue that the exemptions in fact changed the whole conception.

In my view this is a great exaggeration. For it is mainly agreements about minor parameters of action which can be authorized under this law. And in those cases where even the price parameter is included, the cartel may only be authorized if substantial competition prevails. So these cartels can hardly be compared with those agreements which predominated before the Second World War. Those were extended on decisions about prices, supply quantities, production quotas and sometimes even investments in productive capacities. Moreover, they often had been tightened by the establishment of joint purchasing or selling organisations, which covered the whole market. This type of cartel organisation cannot be found anymore in West Germany.

The rather loose agreements which can be authorized under the present law may to some extent cause a price increase of some products and thereby influence the income distribution slightly to the disadvantage of the consumer. But this form of inter-firm organisation is hardly strong enough to coordinate the innovative activities and the investment decisions which in the long run determine the productivity and growth pattern of the economy most effectively.

So the main problem of today's competition policy is not the cartel agreement but the concentration process and especially the merger movement.

## 4. CONTROL OF DOMINANT ENTERPRISES

A second main instrument of German competition policy is the control of the market behavior of market-dominating enterprises. There is no equal instrument in the American antitrust policy. But in some respects the German control may be compared to the prohibition of monopolization according to Section 2 of the Sherman Act.

Since the control of market behavior is limited to enterprises with a market-dominating position, the first step in the control process is the determination of the market position.

In Section 22 the law defines a market-dominating enterprise as one which has 'no competitors' or which is 'not exposed to any substantial competition'. This definition is rather narrow, and it corresponds approximately to the range of firms which under the American jurisdiction may fall under the monopolization prohibition of Section 2 of the Sherman Act.

But together with the introduction of a merger control in 1973 the legal criteria for a market-dominating position were extended. The new criterion now has a double function. It serves as a new intervention threshold for the already-existing control of market behavior and it became the intervention threshold for the newly-introduced merger control.

Since 1973, dominant firms according to Section 22 were not only those which have 'no competitors' or were 'not exposed to any substantial competition'. In addition those firms were included which had 'a paramount market position in relation to their competitors'. Additionally the law enumerates some structural criteria by which a paramount market position can be identified.

Furthermore, the law provides legal presumptions for market domination by a single firm and by oligopolies. These are for a dominant single firm a market share of one-third or more, and for dominant oligopolies a market share of one-half or more for the biggest three firms or a share of two-thirds or more for the biggest five firms. These presumptions are rebuttable. They should make it easier for the Cartel Office to submit proper evidence to the courts.

The second step of the control process is the determination of abusive practices. In their jurisdiction the Federal Cartel Office distinguished two kinds of abusive practices, which are

first, horizontal abuses of market power against competitors, in particular predatory behavior, and
secondly, vertical abuses against customers, especially by raising prices above the competitive level.

Since the third amendment of the law in 1980, these two types of abusive practices are explicitly enumerated in the law together with price discrimination as a third possible type of abusive practice. If we compare this control of dominant firms with the American antitrust law, we reach the following conclusions.

First, the prohibition of horizontal abuses of market power has some similarities with the prohibition of monopolizing according to Section 2 of the Sherman Act:

- On the one hand, the range of firms which fall under the German control is wider, since the legal definition of market domination has been extended in the amendment of the law of 1973. Market domination is now presumed if a firm has a market share of one-third or more, while the American jurisdiction of Section 2 of the Sherman Act does not cover firms with less than 60%.
- On the other hand, it seems that the term monopolization is much more widely defined by American jurisdiction than the term abusive practices in the German one.

Secondly, the prohibition of vertical abuse is essentially a control of market performance. The Cartel Office or the courts have to decide whether a certain price is too high or not. There is no similar kind of control in the American antitrust law at all. This conception is only to be found in the American public utility regulation. But this regulation is much more comprehensive.

Thirdly, the prohibition of price discrimination is determined by Section 26 as well as Section 22 of the German law. Its model of course was the American Robinson Patman Act of 1936. But there is the important difference that the Robinson Patman Act applies to all enterprises while according to Sections 22 and 26 of our law the prohibition of price discrimination is limited to enterprises with a market-dominating position (Section 22) or enterprises on which suppliers or purchasers are dependent (Section 26).

During the last ten years we have had some cases of all types brought to the Federal Supreme Court. The experience was that, while the cases of horizontal abuse of market power could be solved rather satisfactorily, the vertical cases raised many problems[4]. It appeared to be extremely difficult to determine a price level which would have been the result of competition without domination. But such a fiction of a competitive price level is necessary in order to identify the actual price in a market as abusive.

---

[4] Monopolkommission (1975).

## 5. MERGER CONTROL

Until 1973, before the Law Against Restraints of Competition was fundamentally amended, the prohibition of cartel agreements and the behavioral control of market-dominating enterprises were the main instruments of the German competition policy. But these two instruments seemed not to be well balanced. There was a contradiction in the fact that the rather loose organizational form of restraining competition—the cartel agreement—was hit by an outright prohibition, while the much more stable form—the dominant single firm—was only subject to a behavioral control. Probably it was this unequal treatment of alternative ways to restrain competition which contributed much to the strong concentration process during the 1960s.

In 1973, the second and most important amendment of the law was put into effect. It tried to reduce the imbalance, which I mentioned, by two provisions:

First, it expanded the possibilities for cartel agreements between small and medium-size firms, in order to give them better chances of competing against large firms (Section 5a).

And secondly, it established a merger control where all mergers are prohibited which are expected to create or strengthen market-dominating positions (Section 24).

The model for this merger control was Section 7 of the Clayton Act, although the criteria for intervention are slightly different. Today one may say that the merger control has become the most important provision of the law, and it employs a large share of the Cartel Office's resources. Therefore I will discuss this instrument in more detail.

According to Section 24 of the law the Federal Cartel Office has to prohibit a merger if it creates or strengthens a market-dominating position and if there are no improvements of the competitive conditions which outweigh the disadvantages of the market domination.

As I mentioned above, the legal criteria for a market-dominating position were newly defined in the 1973 amendment. Since then dominant firms according to Section 22 were not only those which have 'no competitors' or were 'not exposed to any substantial competition'. In addition, those firms were included which had 'a paramount market position in relation to their competitors', and the law enumerates some structural criteria by which a 'paramount market position' can be identified. Those are 'its share of the market, its financial strength, its access to the supply or sales markets for goods and services, its links with other firms, and the legal or actual barriers to market entry'.

These criteria show very clearly that the merger control was directed against vertical and conglomerate mergers as well as against horizontal ones. With respect to horizontal concentration the law provides in addition some legal presumptions.

## 6. HORIZONTAL MERGERS

If we compare the German merger control with the American one according to Section 7 of the Clayton Act, we find that at the time of its introduction it was probably less severe. This is certainly true if we restrict the comparison to horizontal mergers and if we accept the merger guidelines of the US Department of Justice of 1968 as a yardstick. In these guidelines the critical market share for merging firms was between 8% and 12%. The corresponding German criteria were much higher.

But the 1968 merger guidelines have never been fully accepted by the US courts, and meanwhile they have been replaced by the new guidelines of 1982 and 1984. The new guidelines replaced the concentration ratios by Herfindahl indices and placed the critical values between 1000 and 1800. A simple conversion of Herfindahl indices into concentration ratios, of course, is impossible. Yet a rough comparison is possible, and it was done by two colleagues of mine, Schmidt and Ries[5]. These calculations show that the critical indices of the new merger guidelines are under realistic assumptions in the neighbourhood of the German critical concentration ratios.

But another aspect has to be taken into account before we can draw any conclusions from these comparisons. It is obvious that concentration measures can only be compared if the product markets are equally defined. We can assume that this has been the case before the new merger guidelines were issued. American and German Authorities defined the product markets rather narrowly and included only competitors who actually supplied the product.

---

[5] Schmidt and Ries (1983).

This method was changed in the new American merger guidelines. Now, potential competitors are also included in the market. The relevant criterion is the expectation that the potential competitor would enter the relevant product market if the price level rose by 5%. It is obvious that this kind of measurement will lead to much lower concentration figures, no matter what indices are used. If we take this fact into account, we come to the conclusion that today the German merger control with respect to horizontal mergers is much more rigorous than the present American one.

What strikes Continental Europeans, who are used to a different legal system, most, is the fact that this fundamental change of the American antitrust policy took place without any action of the American legislative bodies and without any change of the law. This would be absolutely impossible under our system of law.

Generally, I have much sympathy for the new approach of the US Justice Department in defining the relevant market. The traditional way very easily leads one to overlook important competitive forces. Recent developments in economic theory as well as the investigation of antitrust cases lead to the conclusion that potential competition may have stronger influences on market performance than the concentration of actual competitors on that single-product market. To some extent these influences are more easily taken into consideration if the relevant market by definition includes potential competitors.

But on the other hand, I have some doubts whether this new concept will always lead to satisfactory results.

At first, it seems rather difficult to me to estimate which firms would enter a relevant market if the price level of that market rose by a certain percentage. This is especially so, in the most common cases, where the potential competitor has to undertake irreversible investments when he enters the market and at the same time does not know what kind of price strategy the incumbent firm will undertake as reaction. So the new critical concentration indices rest on less solid ground than the old ones.

Secondly, actual and potential competitors are evaluated equally. In extreme cases the new procedure may lead to the result that the critical Herfindahl index is not reached, although there is no actual competition in the product market at all. I have doubts whether potential competition can substitute actual competition with respect to all performance criteria.

Thirdly, even if we ignore these two arguments, I still have doubts whether the two concentration thresholds of HHI 1000 and HHI 1800 are low enough with respect to the broad market definition. They are certainly much bigger than the German legal presumptions with respect to the traditional narrow market definition, and until now I did not have the feeling that these were much below the threshold where one had to expect some kind of collusive behavior.

The consideration of the German Monopolies Commission went in a slightly different direction. It holds the opinion that structural criteria of the traditionally defined product market cannot explain the competitive environment of firms completely. It therefore proposed, in addition, the inclusion in the analysis of firms which supply substitute goods, or which, because of similar production techniques, might easily enter the relevant market. The Federal Court of Appeal in its landmark decision on the merger of Guest, Keen and Nettlefold (GKN) and Sachs used the expression that these firms operate in the 'neighbourhood of the relevant market', and the Monopolies Commission has elaborated this concept in particular in its main report of summer 1984[6].

The Commission proposes an analysis of a competitive situation in two steps:

First, the actual, short-run competition on the relevant product market should be analysed in the traditional manner. On this step the market concentration and the product heterogeneity are the most important criteria.

In a second step, the competitive influences, which operate in the long run from the 'neighbourhood of the market', should be taken into consideration. These are especially the elasticity of substitution on the demand side and the flexibility of production on the supply side.

Although the competitive pressures on the behavior of the firms are strongest under conditions of almost homogeneous market competition, for the overall performance of the economy the competitive forces from neighbouring markets and industries are at least as important. The Monopolies Commission therefore has always emphasized that in cases of market domination competition policy should at least

---

[6] Monopolkommission (1984) Chapter VII.

guarantee that suppliers of substitutional products and potential competitors in the neighbourhood of the market stay independent of the dominating firm. With great suspicion do we watch for example the emergence of energy conglomerates and media conglomerates[7].

## 7. NON-HORIZONTAL MERGERS

The differences between the German and American merger control are even greater with respect to non-horizontal mergers, especially towards pure conglomerates[8]. As I understand the American development today, there exists practically no barrier at all any more to conglomerate mergers. And even more surprising, in this respect the new policy seems to find approval from many lawyers and economists who usually are ideologically not located at Lake Michigan.

The German Monopolies Commission evaluated the German Merger control from its beginning. Already in its first main report it pointed out that the Cartel Office should not concentrate its attention on horizontal mergers only, but also investigate carefully the anticompetitive effects of conglomerate mergers[9]. These may be expected from the combination of financial resources of the merging firms or from the combination of high market shares with financial strength. On the basis of a proposal by the Monopolies Commission a fourth amendment of the law was passed in 1980 in which corresponding legal provisions were established (Section 23a).

I think it is justified to call special attention to one of these provisions. According to Section 23a it is presumed that a market domination will arise if two or more firms, with total sales of DM 12 billion or more, merge. This provision violates an important principle of the American antitrust jurisdiction, namely that 'mere size is no offence'. Of course, it does not make mere size an offence. But it acknowledges that mere size and especially the financial strength, which goes together with size, might endanger competition. It might alter the character of the competitive process and discourage smaller firms from competing rigorously against the growing giant.

---

[7] With respect to energy see: Monopolkommission (1976), Chapter IV. With respect to media see: Monopolkommission (1981). See also Kantzenbach (1984).
[8] Monopolkommission (1984) Chapter VIII.
[9] Monopolkommission (1976) Chapter V.

This consideration of financial strength in the determination of market domination is not based on non-economic, political goals. In its motivation it, therefore, differs very much from the bills of Senators Hart and Kennedy and the FTC in the '60s and early '70s, which also had the purpose of inhibiting the formation of giant firms by merger. The Commission did not follow the common American view that limitations to firm size can only be justified by non-economic goals, especially by the ideal of a Jeffersonian middle-class democracy.

It believes that there are also good economic arguments against a paramount concentration of financial resources. These have the property that they can easily be transferred from one branch of the enterprise to another and, therefore, can be used as a strategic instrument in any market in which the enterprise is engaged. They can be used for the development and introduction of new products, for the advertising and subsidizing of certain products, for deterrence of new entry and so on.

Of course this is the old-fashioned 'deep pocket doctrine'[10], and there have been many statements published against it in recent years. In particular only very few cases have been found in which conglomerate enterprises have been convicted for predatory action[11].

But, on the other hand, we know the strategic models of product portfolios, which were first developed by the Boston Consulting Group. Within these models it appears to be the most natural business strategy of conglomerate firms to drag money out of old products with a strong market position (to milk the cash cows), and to use it for the development and marketing of new products (to feed the wild cats). I do not argue that these are abusive practices *per se*. But it seems to be obvious that in this respect a conglomerate multiproduct firm has competitive advantages over single-product firms with much smaller financial resources. These advantages have nothing to do with the production costs or the quality of one product and therefore with the allocative, technical or qualitative efficiency of the relevant market. They may be used to drive smaller competitors out of the market, although that does not seem to be the rule. But the existence of paramount financial strength as such implies a threat to the competitors

---

[10] Edwards (1955), Narver (1967).
[11] Markham (1973).

which induces them to 'behave well', to accept a price leadership for example. Therefore, it seems to be justified to acknowledge the overall size of an enterprise and the financial strength which goes with it as a possible independent cause of market domination.

The German Supreme Court adopted this view with some limitations in the merger case GKN/Sachs. According to this decision the financial resources of one merging firm will strengthen the market dominating position of the other merging firms if the first firm operates in the 'neighbourhood of the market' of the other firm. In other words, the court expects that the profits of one product may be used in another product market if there are some 'commonalities' between these two products. The decision could be interpreted as the acceptance of the deep pocket doctrine in product-extension merger cases.

In the case of Württembergische Metallwarenfabrik (WMF) and Rheinmetall, the court seemed to have dropped the limitations of the former case. It pointed out that the financial resources of one merging firm must generally be expected to be at disposal for competitive action on the markets of the other merging firm. So it seems that the court adopted the deep pocket doctrine also for pure conglomerate mergers.

## 8. MAIN CRITICS OF THE GERMAN POLICY

Until now I have tried to show that the German competition policy comes to quite different conclusions than the present American antitrust policy, although it rests on the same basic philosophy. I want to close with a few remarks about the criticisms which are brought forward against this policy.

The present German competition policy has more similarities with the American policy of the '60s than the present one. One would expect that in such a situation many authors would accept the theory which led to the fundamental change in the American policy. One would expect that such authors would propose similar changes in German competition policy. Surprisingly, this is the case only to a very limited extent. As far as I can see, only the German Association of Manufacturers (Bundesverband der Deutschen Industrie) adopted the theory and the reasoning of the Chicago school without substantial change.

In academic discussions and professional journals another kind of critique gained some attention. Some authors, especially Erich

Hoppmann[12] and Dieter Schmidtchen[13], hold the view that competition policy, because of its very nature, can only pursue one goal, namely freedom of competition. The only criterion for evaluating a competitive process, according to them, is the individual market behavior, whether it impairs the competitive freedom of other firms in the market or not. The authors, in particular, reject the use of market structure and market performance criteria to evaluate competitive processes. Market structures in their view are results rather than causes of market processes. Neither the outcoming market structure nor the common performance criteria justify any conclusion as to whether the underlying market process had been competitive or not. From this point of view many antitrust actions are criticized as interventionistic and directed against the free economic order. These authors are deeply influenced by the thinking of Friedrich von Hayek. Sometimes they call themselves the Neo-Austrian School. One outstanding American author of the school is Israel Kirzner[14].

Although with respect to competitive policy these authors reach quite similar final conclusions, the Neo-Austrian School may not be confused with the Chicago School, with Richard Posner, Harold Demsetz, Yale Brozen and others. Their starting points are quite different. For the Chicago people—and for the present American government policy—economic efficiency is the only goal of antitrust. For the Neo-Austrians efficiency is a collectivistic (holistic) fiction with no expressive power at all. What should be taken into account are only individual goals, and individual freedom is the only possible criterion for a free economic order.

So, from our comparison of American and German competition theory and policy, we reach a rather strange and confusing conclusion.

In America a very passive, almost laissez faire policy is observed, which is justified by its supporters with purely economic reasoning. The critics of this policy are asking for more rigorous antitrust action on general political, in particular Jeffersonian, grounds.

In Germany the situation seems to be quite the contrary. Present competition policy is rather active with growing perfection. Its

---

[12] Hoppmann (1967a, 1967b, 1968)
[13] Schmidtchen (1978).
[14] Kirzner (1973). For a discussion of the different positions in particular see: Gotthold (1981, 1982), Möschel (1981).

orientation is mainly, although not exclusively, economic. The theoretically most influential critics of this policy recommend less policy intervention into the market process, and they justify their recommendations on non-economic, sociophilosophical grounds.

## BIBLIOGRAPHY

Edwards, C. D., 1955 'Conglomerate Bigness as a Source of Power', in: *Business Concentration and Price Policy*, New York/Princeton; NBER.
Edwards, C. D., (1978), 'American and German Policy towards Conduct by Powerful Enterprises: A Comparison', *Antitrust Bulletin*, 23 83ff.
Gotthold, J, (1981), 'Neuere Entwicklungen in der Wettbewerbstheorie', *Zeitschrift für das gesamte Handelsrecht und Wirtschaftsrecht (ZHR)*, 145.
Gotthold, J, (1982), 'Nochmals: Kritische Bemerkungen zur neo-neoliberalen Theorie der Wettbewerbspolitik', *ZHR*, 146.
Hoppmann, E, (1967a), 'Wettbewerb als Norm der Wettbewerbspolitik, *Ordo* 18.
Hoppmann, E, (1967b), 'Workable Competition als wettbewerbspolitisches Konzept', in: *Theoretische und institutionelle Grundlagen der Wirtschaftspolitik*, Festschrift für Theodor Wessels, Berlin.
Hoppmann, E, (1968), 'Zum Problem einer wirtschaftspolitisch praktikablen Definition des Wettbewerbs', in: Schneider, H. -K. (ed.), *Grundlagen der Wettbewerbspolitik*, Schriften des Vereins für Socialpolitik, N. F. **48**, Berlin.
Kantzenbach, E, (1984), '10 Jahre Monopolkommission', *Wirtschaft und Wettbewerb*, **34** No. 1.
Kirzner, I. M. (1973). *Competition and Entrepreneurship*, Chicago; in German: *Wettbewerb und Unternehmertum*, Freiburg.
Markham, J, (1973), *Conglomerate Enterprise and Public Policy*, Boston.
Mestmäcker, E-J and Richter, R (eds), (1980), 'Competition Policy: German and American Experience, A Symposium', *Zeitschrift für die gesamte Staatswissenschaft*, **136**, 3.
Monopolkommission, (1975), 'Anwendung und Möglichkeiten der Misbrauchsaufsicht über marktbeherrschende Unternehmen seit Inkrafttreten der Kartellgesetznovelle', *Sondergutachten I*, Baden-Baden.
Monopolkommission, (1976), 'Mehr Wettbewerb ist möglich', *Hauptgutachten I*, Baden-Baden.
Monopolkommission, (1981), 'Wettbewerbsprobleme bei der Einführung von privatem Hörfunk und Fernsehen', *Sondergutachten II*, Baden-Baden.
Monopolkommission, (1984), 'Ökonomische Kriterien für die Rechtsanwendung', *Hauptgutachten V*, 1982/83, Baden-Baden.
Möschel, W, (1981), 'Kritische Bemerkungen zu kritischen Bemerkungen', *ZHR* 145.
Möschel, W, (1983), *Recht der Wettbewerbsbeschränkungen*, Köln.
Narver, J. C. (1967), *Conglomerate Mergers and Market Competition*, Berkeley.
Schmidt, I, (1973), *US-amerikanische und deutsche Wettbwerbspolitik gegenüber Marktmacht*, Berlin.
Schmidt, I, (1983), 'Different Approaches and Problems in Dealing with Control of Market Power: a Comparison of German, European and US Policy Towards Market-dominating Enterprises', *Antitrust Bulletin*, **28**, 417ff.

Schmidt, I. and Ries, W. (1983), 'Der Hirschmann-Herfindahl-Index (HHI) als wettbewerbspolitisches Instrument in den neuen US-Fusionsrichtlinien 1982, *Wirtschafts und Wettbewerb*, **33**, 178.

Schmidtchen, D, (1978), *Wettbewerbspolitik als Aufgabe: Methodologische und systemtheoretische Grundlagen für eine Neuorientierung*, Baden-Baden.

# Competition Policy in the European Community*

KEN GEORGE
*University College of Swansea*

and ALEXIS JACQUEMIN
*Université Catholique de Louvain, and Advisor to the Forward Studies Unit of the Commission of the European Communities*

## 1. INTRODUCTION

### 1.1. A Brief Review

The European Community's competition policy is laid down in Articles 85 and 86 of The Treaty of Rome (1957).

Article 85(1) states that 'agreements between undertakings, decisions by associations of undertakings and concerted practices which may affect trade between Member States and which have as their object the prevention, restriction or distortion of competition within the Common Market shall be prohibited as incompatible with the Common Market'. There is, however, provision for exemptions including block exemptions as, for instance, in the case of franchising agreements.

Article 86 states that 'any abuse by one or more undertakings of a dominant position within the Common Market or in a substantial part of it shall be prohibited as incompatible with the Common Market in so far as it may affect trade between Member States'.

These basic principles have been translated into an effective policy by various implementing Regulations, notably No. 17 which came

---

* The views expressed in this chapter are personal ones and are in no way attributable to the Commission's thinking on competition issues. The authors are grateful to Mr. Hans Liesner for comments on an earlier draft.

into force in 1962. This Regulation gives exclusive power to the Commission to enforce its policy. For this purpose one of the European Commissioners is given responsibility for the implementation of competition policy and he is assisted by a Directorate-General for competition, DG IV.

There is no article in the Treaty of Rome that deals specifically with mergers, though both Articles 85 and 86 have been used, with mixed success, to deal with merger cases. In 1973 the Commission put forward a draft merger-control proposal to the Council of Ministers but this never succeeded in becoming law. However, in December 1989 the Council accepted a merger regulation which will give the Commission powers for dealing with mergers above a prescribed minimum size in terms of turnover, and which have a 'Community dimension'.

The EC has adopted an administrative approach to competition policy in which the Commission has extensive powers to investigate suspected breaches of the law, to require the ending of anticompetitive practices and to impose penalties on offending parties. With the powers both of prosecution and judge there has to be provision for appeal against the Commission's decisions. This power of review lies with the European Court of Justice.

### 1.2. Policy Objectives

Historically, European countries have viewed the virtues of competition with scepticism. In all the main industrial counties. UK, West Germany, France and Italy, policies of cooperation and government intervention in industry have often been seen as a better way of achieving economic and social objectives than the free play of market forces, although the exact form of these alternative policies have of course varied from country to country. Part of the reason for this scepticism can be traced to the tradition in Europe of placing more emphasis on competition as a process of change rather than as a mechanism which, given certain ideal conditions, will lead to an ideal allocation of resources. Much of the scepticism can thus be traced to the view that competition either will not achieve a desirable objective quickly enough or will do so but at an unacceptably high social cost. This view is indeed reflected in the EC's *First Report on Competition Policy* (1972): 'Even though the operation of market forces is an irreplaceable factor for progress and the most appropriate means of ensuring the best possible distribution of production factors, situations can nevertheless arise when this by itself is not enough to obtain the required results without too much delay and intolerable social tension'.[1]

Nevertheless competition policy is an important part of the general economic and industrial policy of the EC and is indeed becoming more prominent as the completion of the internal market, scheduled for 1992, gathers pace. The emphasis as already mentioned is on competition as a dynamic process. To quote again from the First Report on Competition Policy: 'Competition is the best stimulant of economic activity since it guarantees the widest possible freedom of action to all. An active competition policy makes it easier for the supply and demand structures continually to adjust to technological development. Through the interplay of decentralised decision-making machinery, competition enables enterprises continuously to improve their efficiency which is the *sine qua non* for a steady improvement in living standards'. Efficiency therefore is not to be viewed statically but rather as something that evolves. It manifests itself in the continual emergence of new products, new firms, new industries, new methods of production and organisation and the disappearance of outdated products and processes, all of which are closely bound up with research, technical progress and innovation.[2] It follows that restrictive practices which hinder or obstruct these dynamic processes are to be condemned—hence Article 85(1). At the same time it seems to be accepted that these same processes may lead to the emergence of dominant firms. So it is not the attainment of dominance that is necessarily condemned but any abuse of that dominance should it occur—hence Article 86.

It would appear therefore that EC competition policy is not much concerned with market structure as such. So long as the market mechanism operates unhindered by restrictive practices and untainted by the abuse of market power this would seem to be sufficient. Yet, on this issue the picture is less than clear.

One example that appears to illustrate the Commission's concern for market structure is that of the viability of small and medium-sized enterprises. In the *Seventeenth Report on Competition Policy* (1988)

---

[1] European Economic Commission, *Competition Policy*, First Report, Brussels, April 1972, p. 11.
[2] European Economic Commission, *op.cit.*, pp. 11–12.

the Commission expresses its concern 'to strengthen and preserve small and medium-sized enterprises (SMEs) as an essential element in a healthy environment ... because of the contribution they make to the competitive structure of the market, their flexibility and their dynamism'.[3] Furthermore, although the wording of Article 86 appears to attack the abuse of a dominant position rather than its attainment, the Commission has argued that Article 86 can be used to *preserve* competitive market structures, an interpretation which has been upheld by the Court of Justice.[4] Finally, another clue to the Commission's concern about market structure is the emphasis which it has placed on introducing a regulation that deals specifically with mergers that create dominant positions. The Commission has for some years been pressing for the introduction of a regulation to deal specifically with mergers. Its concern has been heightened with the realisation that the run-up to 1992 may well induce a spate of merger activity which might in some case have markedly anticompetitive intentions or effects.

## 2. HORIZONTAL COLLUSIVE BEHAVIOUR

Article 85(1) prohibits agreements, decisions and concerted practices which have the object of preventing, restricting or distorting competition within the Common Market and which affect trade between member states. The main targets of this prohibition include price-fixing, market-sharing, supply restrictions and the tying of sales. The prohibited agreements are automatically void but can under certain conditions be exempted. The conditions, spelt out in Article 85(3), are:

(i) that the agreement or practice must contribute to improving the production or distribution of goods or to promoting technical or economic progress,
(ii) that consumers get a fair share of the resulting benefits,
(iii) that the restriction is necessary for attainment of the objective,
(iv) that the firms concerned are not thereby enabled to eliminate

---

[3] Commission of the European Communities, *Seventeenth Report on Competition Policy*, Brussels, 1988, p. 29.
[4] This was established in the now famous *Continental Can* case which is discussed later.

competition in respect of a substantial part of the product in question.

The prohibition of Article 85(1) expressed a principle of the European Community (EC). The effectiveness with which this principle is applied depends on the procedures that have emerged for administering the law and the way in which the law has been interpreted in individual cases.

Regulation 17 gave the European Commission considerable powers as the exclusive enforcement agency of EC competition policy. It has wide powers to seize evidence and if it uncovers an infringement of Article 85 the agreement is automatically null and void and heavy fines can be levied on the participating firms. The Commission can initiate investigations where a complaint has been made by an interested party or where the performance of an industry gives reason to suspect an unlawful practice. Firms which wish to continue operating an agreement without threat of discovery and fine have two options open to them: they can apply for 'negative clearance', i.e. a declaration by the Commission that their agreement is not prohibited, or they can notify the Commission of their agreement and seek exemption under 85(3). Failure to notify does not condemn an agreement but it cannot be exempted.

If the Commission finds fault with an agreement it can terminate that agreement by issuing a formal decision. It may alternatively make an informal recommendation to the firms concerned to voluntarily terminate an agreement or to modify it so that it is no longer objectionable. Where an exemption or negative clearance is granted conditions may be attached to the decision so that it may be revoked if the conditions are not adhered to.

Since 1970 it has been established that an agreement between firms is not prohibited by Article 85 if its effect on competition and trade between member states is insignificant, insignificance being defined in terms of the combined market share of participating firms in the product concerned or in terms of aggregate turnover.

Concerning the possibility of exemption under Article 85(3), this requires a trade-off to be made between an increase in market power on the one hand and possible efficiency gains on the other: i.e. some collusive behaviour restricting competition in a non-minor way may be exempted because of sufficient beneficial effects. This possibility of an 'efficiency defence' is in contrast to the US tradition, although recently the American enforcement agencies have moved in that direction. Such a defence has its counterpart in Williamson's formula for measuring the respective sizes of cost-savings and surplus reduction due to a restriction of output. However, this 'naive', static, partial-equilibrium model, with its cost-benefit analysis limited to two-dimensional terms, requires a number of qualifications which strongly reduce its operationality. These include matters of timing, non-price competition, X-inefficiency, income distribution effects, second-best considerations, as well as the inference and enforcement expenses needed to prove the existence of benefits. What is in fact suggested by the model is the difficulty of identifying at all precisely the efficiency consequences of a business conduct and of advocating finely-tuned optimal antitrust rules. The conditions of Article 85(3) therefore do not rest on a strict and unambiguous welfare analysis and as a result there is a danger that decisions could be the outcome of political compromises between conflicting and incommensurable values. This danger is reduced in two ways.

The first and most obvious one is that even when the Commission can be persuaded that an agreement makes a valuable contribution to one of the relevant economic goals mentioned in 85(3) it still has to be convinced that the agreement does not allow the firms concerned to eliminate competition in respect of a substantial part of the products in question. No matter how beneficial the effects of an agreement therefore, if it eliminates competition for a substantial part of the product it will be condemned. To be sure, this still leaves room for interpreting what constitutes substantial; but the wording of the legislation certainly allows the Commission to adopt a robust approach to maintaining competition.

The second way in which the burden of the trade-off has been reduced is that Article 85(3) has been implemented not so much on a case-by-case basis but by granting block exemptions dealing with important types of agreements (e.g. agreements relating to specialisation, R&D and patent licensing) for which there exists a presumption that a situation of market failure can exist. This system of exempting classes of agreement from the notification requirement avoids the necessity for a detailed scrutiny of each conduct. Rather, it creates codes of conduct that can increase the credibility of the policy and limit the discretionary power involved in the Article. At the same time it preserves the Article's valuable message that antitrust policy must be sensitive to possible efficiency gains and that there are circumstances when cooperative behaviour can restrict competition in a non-negligible way and still produce socially desirable results.

From the outset the Commission has dealt firmly with agreements involving price-fixing, quantity restrictions or market-sharing arrangements. For all cartels, national as well as international, which affect trade flows, the position with respect to these practices has been one of near *per se* prohibition. In the early years of enforcement the attitude towards joint sales agencies was more relaxed. Of the fourteen cases decided up to the end of 1973 two were given negative clearance, six were accepted following modification and six were condemned and had to be abandoned. They were concentrated in the producer good sector and involved relatively homogeneous, capital intensive products, such as certain steel products, fertilizers and cement, where competition was liable to be fierce during times of depressed demand. Lower costs following from 'orderly marketing', rationalisation of production, and coordination of investments in order to avoid excess capacity were the kinds of argument put forward to support these selling syndicates. Experience however proved otherwise. Prices were often found to be aligned to the highest cost producers and there was little evidence of flexibility in the face of demand fluctuations. Consequently by the early 1970s the Commission's attitude to sales agencies had hardened. By this time the position with respect to formal cartel arrangements had become clear: 'horizontal cartels will only be allowed in those cases where their market share is unimportant'.[5]

But article 85 covers much more than formal cartel arrangements. It also embraces concerted practices, 'open-price' or information agreements, and the use of industrial property rights to restrain competition.

With formal horizontal cartels being treated as virtually illegal *per se* it can be expected that this would give rise to more subtle collusive arrangements. An early case was that of price collusion between the manufacturers of dyestuffs in the years 1964–67. Parallel pricing was achieved by means of one or more of the producers announcing a price

---

[5] For a review of the early years of policy towards restrictive agreements see de Jong, H. W. (1975), 'EEC Competition Policy Towards Restrictive Agreements', in George, K. D. and Joll, C., (eds) *Competition Policy in the UK and EEC*, Cambridge University Press, 1975.

increase which was then followed immediately or after a short time-lag by other suppliers. The companies' defence was that the parallel price movements were not the result of concerted behaviour but the inevitable outcome of the oligopolistic structure of the market. A firm which failed to follow a price rise could not, it was argued, have gained a competitive advantage because its competitors would revise prices downward or withdraw their announced increases with the result that the profitability of all would be impaired. This might correspond to cases analysed in the economic literature on 'non-cooperative collusive behaviour' in repeated games: when firms are few and their objectives are similar they can calculate and move to collusive equilibria without collaborating, solely on the basis of their individual choices.[6] The Commission and the Court did not accept this type of argument but concluded that the firms were guilty of concerted action designed to restrict competition.[7]

The Commission has also condemned some information agreements. One such case was that of the European glass packages manufacturers which was decided in 1974. A group of 26 producers had operated a 'fair competition rules' agreement in which certain practices such as price-cutting to gain a competitive advantage were deemed to be unfair. The agreement involved, amongst other things, the exchange of information on prices, a scheme which was designed to detect price-cutting and which, if discovered, meant the possibility of a fine of at least 30 percent of sales in the market where the price-cutting occurred. The Commission declared the agreement illegal making the obvious point that the agreement which according to the producers was designed to ensure fair competition in fact eliminated it.

In 1987 the Commission imposed its first fine in respect of an exchange of information agreement; this being the case of the three major European producers of fatty acids—Unilever, Henkel and Oleofina who together accounted for 60 percent of the market. The Commission argued that by exchanging information about sales they were able to monitor each other's activities which tended to militate

---

[6] See Jacquemin, A. and Slade, M. (1989), 'Cartels, Collusion and Horizontal Mergers, Chapter 7 in *Handbook of Industrial Organisation*, Schmalensee R. and Willig, R. (eds), Amsterdam: North Holland, 1989.

[7] For further details of this case see de Jong, H. W., *op.cit.*

against the adoption of competitive behaviour and to stabilise market shares.[8]

Industrial property rights such as trademarks, patents, copyrights and the licences which can be given for such rights may serve as the instruments for the separation of markets and may permit firms to pursue anticompetitive practices both nationally and internationally. In a number of cases involving property rights the Commission and the Court have established that these rights can not be used to restrain competition. Most significantly perhaps it has ruled that trademarks, licensing agreements and copyright law may not be used to stop parallel imports, i.e. goods imported through channels other than the distributive network set up by the manufacturer of the goods.[9]

### 2.1. Cooperative Agreements and State Aids

While the Commission has taken a consistently hard line in relation to classic anticompetitive violations such as price-fixing and market-sharing agreements it recognises the fact that the unfettered play of competitive forces will not always give the best results, particularly in relation to dynamic adjustment to changes in demand and supply conditions. Thus there is a role, within carefully defined limits, for both state aid and cooperative agreement between firms. These are justified on the grounds that free-market forces may work too slowly in bringing about a desired result; they may involve unacceptably high social costs, and they do not always guarantee that the most efficient firms will survive. On the basis of such arguments the Commission has in a number of cases looked sympathetically at agreements and state aids aimed at reducing capacity in industries with a structural over-capacity problem, and has also viewed in a favourable light several cases of state aid and cooperative agreements aimed at fostering research and development and the strengthening of key industries of the future.

It would not be correct to say that the Commission regards cooperation and state aids simply as *alternative* adjustment mechanisms although in some cases this may be the case. It also sees these policies as

---

[8] See Commission of the European Communities, *Sixteenth Report on Competition Policy*, Brussels, 1987, p. 57.
[9] See de Jong, H. W., *op.cit.*

very often *reinforcing* competition, an attitude which is most clearly evident in its approach, discussed below, to small and medium-sized enterprises.

## 2.2. Industry Restructuring

Although the Commission recognises that capacity adjustments to long-run changes in demand are, in ordinary circumstances, the concern of individual firms it also argues that 'within a given crisis-struck industry, economic circumstances do not necessarily guarantee a reduction of the least profitable surplus capacity. Undertakings which have failed to make the necessary adjustments may have their losses offset within their groups ... Against this background, and in order to combat the structural problems of individual sectors, the Commission may be able to condone agreements in restraint of competition which relate to a sector as a whole ...'.[10] Condonement of such agreements is always subject to conditions which are, broadly, that the criteria laid out in 85(3) have to be satisfied. Importance is also attached to the benefit that a restructuring agreement may offer in terms of reducing the social costs of plant closures especially by way of the retraining and redeployment of workers made redundant. In addition the Commission will insist on a detailed and binding programme of rationalisation which has to be achieved within a specified period of time. During the depressed years of the early 1980s the Commission dealt with a number of joint measures to reduce overcapacity. The first formal decisions were taken in 1984, one of the earliest being the exemption accorded to an agreement between the ten largest European synthetic fibre manufacturers to close 18 percent of their production capacity for six types of synthetic textiles.[11]

As an alternative to sectoral agreements the Commission has also cleared agreements between a small number of firms providing for reciprocal arrangements which would enable capacity to be reduced and lower costs to be achieved as a result of increased specialisation. One example of this dimension of policy was the Commission's

---

[10] Commission of the European Communities, *Twelfth Report on Competition Policy*, Brussels, 1983, p. 43.
[11] Commission of the European Communities, *Fourteenth Report on Competition Policy*, Brussels, 1985, p. 70.

acceptance in 1984 of a bilateral agreement between two British companies (ICI and BP Chemicals) which involved an exchange of production facilities, technology and goodwill in the UK, so that one company could specialise in PVC and the other in low density polyethylene. The Commission accepted the agreement because the advantages of more efficient production would be achieved while maintaining effective competition, both in the UK and the wider EC market.

Much the same approach is adopted for state aids, which have been, and still are, a major problem facing the Commission. Generally, they delay the reallocation of resources and because of their unequal incidence across the Community distort trade between member states. Yet in certain areas state assistance is accepted as having net beneficial effects. One such area is aid to accelerate structural change in industries which face long-term structural problems, and to ease the social problems associated with this change. Once again, however, the Commission invariably attaches conditions to the blessings that it gives in this area. For instance, its authorisation of aid to the European steel industry in 1983 was made subject to two major conditions, namely a firm timetable for the implementation of cuts in capacity and a demonstration that the aided firms would be financially viable at the end of the aid programme.

It could of course be argued that changes of the kind that have been discussed could be safely left to market forces and that this would result in a more efficient outcome. If, for instance, there are major benefits to be achieved from specialisation this must surely be recognised by at least some firms in the market who will seek to achieve them. However, the competitive process may be very prolonged and the financial risks, especially in a stagnant or declining market, may be such that firms are unwilling to accept them. Unfettered competition may result in all firms suffering cuts in profits with little benefit in terms of increased specialisation.[12] On balance, the Commission has adopted a sensible and realistic approach to policy in this area.

### 2.3. R&D and New Industries

Just as the Commission has recognised the special problems of industries facing major structural weaknesses so it has been concerned to

---

[12] See, for instance, George, K. D. (1987), 'Rationalisation of Industry' in Eatwell J., Milgate M., and Newman P. (eds) *The New Palgrave: A Dictionary of Economics*, Vol. 4; London: Macmillan, Press 1987.

create favourable conditions for R&D and the development or new industries. These matters are viewed not only in the light of competitive conditions within the Common Market but also in relation to the competitive threat posed by countries outside the EC, notably Japan and the US.

The Commission has been guided by the view that because of the resources needed and the risks involved many R&D programmes would not take place at all in the absence of cooperation. In many cases, also, the diffusion of new ideas and their transformation into new products and processes will be facilitated by joint efforts. These views correspond to well known economic arguments that have been recently confirmed in oligopoly models.[13] The amount of R&D undertaken by private firms and the diffusion of new knowledge may be socially inefficient over a broad range of market structure. Cooperative R&D can then be viewed as a means of simultaneously internalising the externalities created by significant R&D spillovers, hence strengthening incentives and limiting wasteful duplication by providing an efficient sharing of information among firms. Further, but more problematic, is the view that there are fundamental limits on the ability to protect intellectual property given that scientific knowledge has many of the characteristics of a public good. Hence the full exploitation of the results of cooperative research may require cooperation at the level of production and even distribution. If firms are prevented from such joint exploitation and if it is anticipated that the benefits of cooperative R&D will be quickly dissipated through intense product market competition, firms will be tempted either to avoid R&D cooperation and to maintain wasteful competition in the pre-innovation market or to use their cooperation to unduly limit the amount of R&D activity. If this is true a regulation of R&D cooperation that excludes cooperation at the level of final markets could discourage or destabilise some valuable agreements. On the other hand, allowing an extension of cooperation from R&D to manufacturing and distribution tends to encourage or underpin collusive behaviour. This is precisely the dilemma faced by the European authorities and the block exemption regulation on R&D

---

[13] See Katz, M. (1986): An Analysis of Cooperative Research and Development, *Rand Journal of Economics*, **17**; and d'Aspremont Q, and Jacquemin A. (December 1988), 'Cooperative and Non-Cooperative R&D in Duopoly with Spillovers, *American Economic Review*. For a general analysis see, Jacquemin, A. (1988) Collusive Behaviour, R&D and European Economic Policy, *European Economic Review*, **32**.

agreements which came into force in March 1985 expresses the compromise which has been adopted.[14] The regulation leaves intact the 1968 Notice on Cooperation between Enterprises which state that cooperation agreements relating only to R&D normally do not fall under Article 85(1), but extends this favourable treatment to R&D agreements which also provide for joint manufacturing and joint licensing to third parties (but not joint distribution and selling). The block exemption is only applicable if all parties have access to the results; if, when there is a joint exploitation, each party is free to exploit independently the results; and if in the case of specialisation in the manufacture of the new or improved products, each party has the right to distribute those products.

In order to ensure the maintenance of effective competition the block exemption sets out different conditions according to the competitive positions of the parties. Thus, for instance, if the parties to the agreement are not competitors the exemption applies irrespective of market shares; but where they are, the exemption applies only if their combined production of the products which may be improved or replaced by the results of R&D does not exceed 20 percent of the market. Individual exemptions may however be given to firms with high market shares if the Commission is convinced that cooperation is needed to meet the costs and risks involved in R&D and that effective competition is not endangered. For instance, in 1978 a joint R&D agreement between Beecham and Parke Davis was given a 10-year exemption from the ban on restrictive practices, the conclusive point being the cost involved, even for large pharmaceutical companies, in developing a completely new product in this market.

The Commission's favourable attitude towards R&D is even more evident in the case of State aids, where it has adopted a general presumption in favour of such aid. This stance is clearly stated in the Sixteenth Report on Competition Policy for the year 1986. This general presumption is justified on several grounds notably: (a) 'the risks attached to R&D and the long pay-back periods which may be involved so that the activity would often not take place without aid'; (b) the fact that 'aid for R&D by its very nature may be less prone to distort trade between member states than aid for investment which has a direct

---

[14] See, Commission of the European Communities, *Fourteenth Report on Competition Policy*, Brussels, 1985, pp. 37–39.

impact on production capacity and output'; and (c) because 'R&D is necessary to provide a constant stream of new products so as to sustain growth, prosperity and worldwide competitiveness of Community industry'. At the same time the Commission recognises that aid for R&D is now one of the main forms of government support for industry and there is a danger that it may be used as a subterfuge for other forms of state support.

### 2.4. Small and Medium-Sized Enterprises

An important strand running through the Commission's thinking in recent years has been the conviction that special attention should be given to small and medium-sized firms (SMEs), and that as far as possible the constraints of competition policy should fall on them lightly, if at all. This growing interest in SMEs has developed following the massive loss of employment in the large firm sector—resulting from the decline of industries such as steel and heavy engineering and the advance of automation in industries such as motor vehicles and household electrical goods.

The Commission justifies its posture towards SMEs on the grounds 'that such firms are an essential and major component of a healthy, competitive environment, in view of their contribution to the competitive structure of the market, their flexibility and their role in technical evolution'.[15] The Commission also recognises that SMEs may often be at a disadvantage compared to large firms in their ability to raise funds, to enjoy the advantages of economies of scale, and to get access to specialised services. In markets characterised by dominance, SMEs may also be the victims of predatory behaviour.

The administration of competition policy attempts to correct these imbalances in a number of ways. First, the competitive restrictions of a large number of small firms are exempt from Article 85(1) because their effects are insignificant. Second, some of the block exemptions are of particular relevance to SMEs. In particular the block exemption on specialisation agreements applies only when certain combined market-share and turnover thresholds (currently 20 percent and 500

---

[15] Commission of the European Communities, *Fifteenth Report on Competition Policy*, Brussels, 1986, p. 29.

million ECU) are not exceeded, and the R&D block exemption provides that competing undertakings which together have a market share not exceeding 20 percent may engage in joint R&D and joint exploitation of the results. Third, the Commission looks favourably on state aid for SMEs in the form, for instance, of subsidised loans, financial guarantees, R&D support and management assistance. Indeed, a consideration which will weigh heavily with the Commission in its deliberations on aid to a specific industry is whether such aid discriminates in favour of or against SMEs. Finally, the Commission has the full force of Article 86 to prevent SMEs suffering competitive disadvantage as a result of the anticompetitive behaviour of dominant firms.[16]

### 2.5. Concluding Comments

The approach to horizontal agreements which has evolved is one of balancing the merits and demerits of competition and cooperation. The classical anticompetitive behaviour of cartels such as price-fixing, production quotas, and market-sharing are invariably condemned. On the other hand the Commission recognises the value of many forms of cooperation especially where SMEs are involved. Although the general approach appears eminently sensible this is not to say that there is no criticism to be made of the way in which policy has evolved and of its administration and effectiveness. There are indeed still some areas of concern.

A major problem is of course the sheer task facing the Commission in ferreting out significant restrictions. Although notification of agreements is the only way of gaining exemption, and may also lead to immunity from retrospective fines, these considerations have not proved a sufficiently strong inducement to notify. It has been estimated that possibly over 50 percent of agreements prohibited by Article 85(1) have not been notified to the Commission. Firms are likely to be particularly reluctant to notify those agreements least likely to be exempted and will take a chance that the overworked authorities won't discover their particular case and that, if discovered and brought to account,

---

[16] See, for instance, the *ECS/AK20* and *Eurofix-Bauco/Hilti* cases reported in the 15th and 17th Reports on Competition Policy.

any fines imposed will fall short of the benefits that have been gained. In recent years, for example, Europe's leading chemical companies have on two occasions been caught operating illicit cartels. In 1988 the Commission reported that 15 major petrochemical producers had operated a market sharing and price-fixing cartel from the end of 1977 to at least November 1983. The policy of the cartel was decided at regular meetings which set target prices and annual quotas. And from the mid-1970s to at least the end of 1984 23 chemical companies were found to have operated a price-fixing and quota scheme in the European market for PVC and low-density polyethylene, for which the Commission fixed a fine of ECU 60m (£92m.), the largest ever imposed on an illegal cartel.[17] Another recent case is the raid by the Commission on the offices of ten of the ECs leading cement producers in West Germany, France, Belgium and Italy. The Commission is believed to have discovered evidence of price-fixing, and market-sharing.[18] These and other recent examples demonstrate the Commission's determination to wage war on illicit cartels. They also underline the persistence with which firms in certain industries resort to cartel arrangements even when confronted with the possibility of heavy fines. The problem is most evident in those cases where, because of a relatively homogeneous product, price-cutting is a particularly effective form of competition and where the short-run scope for price competition is substantial due to the presence of heavy fixed costs. It is the very effectiveness of price competition, however, which gives firms the incentive to avoid it, and it is evident that they are prepared to go to great lengths to do so.

The Commission's task in attacking cartels is not made easier by the differences between member states in the emphasis placed on the importance of maintaining competition and also in industry-government relationships. In a review of EC policy towards restrictive practices in the mid-1970s one observer felt compelled to ask: 'Why is it that so many cartel decisions have related to Benelux (and to a lesser extent West German) firms whereas French and Italian firms have been involved in (comparatively) few cases?'[19] The answer which the reader is invited to arrive at is not that cartels are less prevalent in France and

---

[17] See, Commission of the European Communities, *Sixteenth Report on Competition Policy*, Brussels 1987, p. 57.
[18] Reported in *The Financial Times*, April 27 1989.
[19] de Jong, H. W. *op.cit.*

Italy but that the French and Italian governments have been less assiduous than others in enforcing EC policy. During the 1980s there has been a general increase in emphasis on competition policy across the Community but the problem of unequal incidence remains an important one. Another aspect of the same problem are the large differences between member states in public ownership and in relationships generally between industrial enterprises, financial institutions, and the state. State ownership of industry is higher for example in Italy than in the rest of the Community and this makes it difficult for competition policy to have the same impact there. A comparison of the UK with Germany shows that in the latter the banks are more closely involved in industry through ownership of shares and bankers' seats on the supervisory boards of companies. This may make it easier to 'coordinate' business activities without having to resort to any formal arrangements. On state aids generally the Commission is well aware of the scale of the problem which was particularly evident during the recession years of the early 1980s. The Commission has had to adopt a policy of ordering the recovery of aid granted illegally, and because of the disregard by member states of their obligation to give prior notification of aid proposals the policy was extended in 1986 to aid which has been illegally given on procedural grounds only. Even so, the whole area of government–industry relationships and differences that exist between member states is a problem area for EC competition policy.

Finally, although the general thrust of the Commission's thinking on small and medium-sized enterprises has much to commend it, it is arguable at least that policy has become too permissive. For instance, the thresholds set for the block exemptions for specialisation and R&D agreements seem generous—is it sensible to define two firms with a 20 percent combined share of the whole EC market as SMEs? This favourable treatment may in part be explained by the enormity of the Commission's task—a generous definition of SMEs and consequent wide application of the block exemptions is needed so that the Commission can concentrate its available resources on the most serious cases. Similarly, with state aids, these are so numerous, widespread and varied in form that the Commission at best can only hope to deal with those which are likely to cause the most serious competitive distortions. In this area in particular it also has to take account of political realities which make a liberal granting of exemptions virtually inevitable.

## 3. VERTICAL ARRANGEMENTS

Vertical relationships between firms have from the outset been a central part of European competition policy. The Commission has always recognised the legitimate interests that producers have in choosing those channels and methods of distribution which they regard as best suited to the effective marketing of their products. At the same time it has been mindful of the possible anticompetitive effects of certain vertical arrangements especially those of market segmentation. The policy which has evolved may be straightforwardly described as one that permits, indeed even encourages, vertical relationships so long as certain basic conditions relating to the maintenance of effective competition are preserved. The encouragement comes in particular, though not exclusively, in relation to SMEs which are found in large numbers in the distribution sector. The main facets of EC policy towards vertical arrangements concern exclusive distribution and exclusive purchasing agreements; selective distribution and franchising; and discounts to customers. These are now considered in turn.

### 3.1. Exclusive distribution

The Commission has always taken a firm stand against collective exclusive-dealing agreements concluded between groups of manufacturers/importers and distributors, on the grounds that such agreements segment the common market into its constituent national parts and thereby restrict trade between the member states. What follows therefore relates to agreements between a single producer and his distributors. Individual exclusive-dealing agreements pose problems in that the assignment of exclusive sales territories may partition the market at the distribution level and result in the elimination of *intra-brand* competition. The first important test case in this area was *Grundig–Consten* in 1964. Grundig, a German manufacturer had given Consten, a French dealer, exclusive selling rights of its products in France, and also prohibited its non-French dealers from exporting to France. As a result price levels in France were between 20 and 50 percent higher than those in West Germany. The Commission's decision (upheld by the Court in 1966) was that the agreement was an infringement of Article 85. The key factor in this case, which became the foundation of policy towards exclusive dealing agreements, was

Grundig's prohibition on parallel imports thus giving its appointed dealers complete protection from intra-brand competition.[20]

Not surprisingly, therefore, the block exemption issued by the Commission in 1967 allowed bilateral exclusive-dealing agreements subject to certain conditions laid down in Regulation 67/67, especially one stipulating that parallel imports shall not be prohibited. Though not specifically mentioned, the Regulation excluded reciprocal exclusive-distribution agreements between competing manufacturers, an exclusion which was made explicit when the Regulation was replaced in 1983 by Regulation 1983/83. The latter also excluded non-reciprocal agreements of this type, although an exception is made for cases in which one or both of the participating undertakings has a turnover below a prescribed limit (currently 100 million ECU). The new Regulation also increased the scope of exclusive dealing agreements by allowing manufacturers a broader choice in the area allotted to their exclusive distributor, which can now cover the whole of the Community.

The danger that exclusive distribution may lead to market partitioning has been met by an insistence that no barriers be placed in the way of parallel imports. The Commission has shown itself prepared to allow temporary territorial protection in the case of firms which have a small market share if that protection helps the firms to penetrate new markets.[21] However this defence is not available to large firms; as is evident from the *Distillers* case. In the 1970s the Distillers Company Limited (DCL) adopted a price structure designed to protect their sole distributors on the continent where they had a small share of the spirits market. UK wholesalers were not allowed to resell whisky to the Continent except at a price per case £5 higher than that applicable to the UK market. This price differential, DCL argued, was needed by their sole distributors for the effective promotion and marketing of their product on the Continent where brand promotion was more important than price competition in expanding market share. However, the Commission condemned DCL's marketing arrangements because they impeded parallel imports thus eliminating competition in the market for Scotch whisky.[22]

---

[20] Commission decisions of 23rd September 1964 (OJ No. 161, 20th October 1964, p. 254–64).

[21] As, for instance, in the *Trans Ocean Marine Paint Decision* of 1967.

[22] For a discussion of this case see George, K. D. and Joll, C. (March 1978), 'EEC Competition Policy', in *The Three Banks Review*.

## 3.2. Exclusive purchasing

Whereas exclusive distribution endangers *intra*-brand competition exclusive-purchasing arrangements may make access to the market more difficult for competing manufacturers thus jeopardising *inter*-brand competition. This is particularly true when the exclusive-purchasing agreement is part of a network of similar contracts covering those outlets which account for a high proportion of sales.

Exclusive-purchasing arrangements are covered by Regulation 1984/83 (previously they had been covered along with exclusive-dealing arrangements under Regulation 67/67). The underlying principle is that the problem of market foreclosure is linked to the duration and scope of the exclusive-purchasing agreement and the commercial freedom accorded to the resellers. Accordingly the Regulation places a limit on the maximum duration of an agreement. Apart from beer and petrol, where special rules apply, the maximum period is 5 years (renewable). Because beer supply and service-station agreements often entail a substantial financial commitment on the part of the supplier the duration of a tie is 10 years, or longer if a tenancy agreement exceeding 10 years is involved. The Regulation stipulates that the range of tied products must be limited to those which are related technically or by commercial usage. For example, for beer-supply agreements no other products can be included in the tie except other drinks and where a tenancy agreement is involved the tenant is free to use alternative sources of supply for drinks other than beer if they can be obtained on more advantageous terms. (Early in 1989 the Commission announced that it would review tying arrangements for beer as part of a general examination of the EC beer market.)

## 3.3. Selective distribution and franchising

In a wide range of products—e.g. motor vehicles, photographic products, household electric appliances—it is common for suppliers to restrict their outlets to dealers who have suitable trading premises and who are adequately equipped to provide a high level of after-sales service. The Commission has always recognised the legitimate interests that firms have in protecting their reputation in this way. However, there are dangers in that the supplier may weaken competition by restricting the number of qualified dealers and may force them into other obligations relating to such matters as purchases, stock levels, and resale prices. To ensure that selective-distribution systems are not abused, the Court decided in 1977 that the requirements which a supplier makes of his resellers must be laid down uniformly and must not be applied in a discriminatory fashion. Furthermore selective-distribution systems can not be used as a vehicle for restricting or distorting competition.

In the *AEG-Telefunken* case the company claimed that it applied its selective-distribution system non-discriminatorily, but the Commission's conclusions (upheld by the Court) were that the firm had made approval of new dealers dependent on other criteria such as agreeing to charge certain prices, or the protection of existing dealers from further competition, and that it had attempted directly or indirectly to dictate resale prices. It had used its distribution system to keep prices high and to discriminate against new marketing channels, in particular department stores and hypermarkets. The Court recognised that selective-distribution systems tend to blunt price competition but this may be justified by increased competition in quality of service. However the Court made it clear that suppliers are not allowed to maintain high prices by excluding qualified dealers, or to discriminate against firms which have introduced new marketing channels so long as they meet the minimum trading standards laid down by the supplier.[23]

Franchising may be considered as an extreme form of selective distribution in which the franchisor imposes upon the franchisee uniform business methods conducted in premises of uniform appearance as a means of exploiting an intangible property right such as know-how or a trade mark. This method of distribution helps to integrate the European market by facilitating the development of cross-frontier distribution networks. Because of this and because also the firms involved are often comparatively small, the Commission announced in 1986 its intention of introducing a block exemption to cover franchises in the distribution and services sectors. The draft regulation specifies the conditions which have to be fulfilled for the exemption to be applicable, such as the possibility of inter-franchisee sales, and practices

---

[23] Commission of the European Communities, *Eleventh Report on Competition Policy*, Brussels, 1981, p. 66.

which will disqualify a franchising agreement, such as enforcement of resale price maintenance.[24]

### 3.4. Discounts

Discounts off list prices can be pro- or anticompetitive depending on the form which the discounting takes and the firm(s) involved. Most obviously anticompetitive is the case where an association of domestic producers have, as part of a cartel agreement, an aggregate rebate scheme which give dealers an incentive to buy from them rather than from foreign sources. These arrangements would certainly be found to be illegal under Article 85.

The Commission also takes an unfavourable view of such rebates when they are operated by a dominant firm because they tend to restrict the purchaser's scope for choice and raise entry barriers. Thus in the *Michelin Nederland* case (1981) the company applied a discount system consisting mainly in fixing a sales target such that the share of Michelin tyres in a dealers' total sales network either increased or at least remained stable. The Commission found that this policy of collective discounts not only had the effect of tying dealers to the company but also resulted in discrimination between such dealers. The company was found to have abused its dominant position.[25]

More controversial perhaps is the stance which the Commission has developed in relation to price alignment clauses—i.e. meeting a lower price which a customer has been offered by another supplier. In the UK this policy of 'never knowingly undersold', first popularised by the John Lewis retail group, has been generally regarded as beneficial. It benefits customers by reducing search costs and the supplier by strengthening customer attachments thus adding greater certainty to sales volume which in turn leads to more efficient production scheduling.[26] The same considerations would seem to apply in the

---

[24] Commission of the European Communities, *Seventeenth Report on Competition Policy*, Brussels, 1988, pp. 42–43.
[25] Commission of the European Communities, *Eleventh Report on Competition Policy*, Brussels, 1981, p. 67.
[26] See Okun, A. M. *Prices and Quantities: A Macroeconomic Anaylysis*, Oxford: Basil Blackwell. Ch. 4.

relationship between manufacturers and dealers. However the Commission is not prepared to accept that these arrangements are beneficial when operated by dominant firms. Rather it views price alignment clauses as aggravating the abuse of dominance because they automatically provide the manufacturer with full information of the state of the market and on the alternatives open to, and the action of, its competitors, which is of great value for its own market strategy.[27]

## 4. MONOPOLY AND THE ABUSE OF MARKET POWER

Community policy on monopoly is contained in Article 86 of the Treaty of Rome which states that 'any abuse by one or more undertakings of a dominant position within the Common Market or in a substantial part of it shall be prohibited as incompatible with the Common Market in so far as it may affect trade between Member States'. Since policy is directed against firms that have a dominant position this requires that the relevant market be established and that the meaning of dominance be clarified. Both have caused problems in the evolution of Community policy.

### 4.1. Defining the Market

It was the problem of defining the market that resulted in the Commission losing its case against *Continental Can*, brought in 1971, the company winning an appeal on the grounds that the Commission had defined the product market too narrowly and thus had underestimated competition from substitute products. (This was in fact the Commission's first attempt to apply Article 86 to a merger and is discussed further in Section 5). Determining the geographical extent of the market may also be problematic. Of necessity, decisions here are on an *ad hoc* basis; market boundaries may be found to lie within a member country or extend to the whole of the Community or indeed beyond it.

---

[27] See the Commission of the European Communities, *Ninth Report on Competition Policy*, Brussels, 1979.

## 4.2. The Existence of Dominance

In determining the existence of dominance the Commission and the Court have gone beyond the structural characteristics of a market. Dominance depends not only on structure but also on ability to have substantial influence on the market, i.e. power to act independently of competitive pressures. Indeed, in a 1965 memorandum the Commission asserted that dominance is 'primarily a matter of economic potency, or the ability to exert on the operation of the market an influence that is substantial . . .'[28] With this approach, however, dominance can only be proved by the *existence* of abusive conduct and where the latter is absent so is dominance. There is no such thing, therefore, as a good dominant firm.

This unsatisfactory position was to some extent resolved by the Court's judgements in the *United Brands* (1978) and *Hoffman-La Roche* (1979) cases. From these cases it is clear that extremely large market shares, 80 percent in the Hoffman-La Roche case, of themselves constitute evidence of dominance. Dominance can generally be said to exist once a market share of the order of 40-50 percent is reached, but this does not automatically give control so that other factors must be taken into account. These include the structure of the undertaking (e.g. its degree of vertical integration, and control over the distribution process) as well as the structure of the market—i.e. the strength and number of competitors, the degree of potential competition and, especially it would seem, the market share of firms ranked immediately below the leader. These judgments also refer to aspects of conduct and performance as factors constituting evidence in support of a finding of dominance—such as advertising expenditure, leadership in technical knowledge and success in defending market share even when prices are higher that those charged by competitors.

In the *Hoffman-La Roche* case the Court concluded that the company had a dominant position in groups of vitamins where its market share was 45-80 percent because these were way ahead of the market share achieved by its nearest competitor. The firm's performance and

---

[28] For an early analysis of EC policy towards dominant firms see Jacquemin, A. (1975), 'Abuse of Dominant Position and Changing European Industrial Stucture', in George K. D. and Joll, C. *Competition Policy in the UK and EEC*, Cambridge University Press, 1975.

results obtained from its policy were regarded as being of only secondary evidential value. Interestingly also in this case, particularly in the context of the new draft regulation on mergers, the Court denied the relevance of overall company size in determining dominance.

Similar considerations applied to the *Michelin Nederland* case (1981). The company's dominant position was found to result first and foremost from the fact that it held 60 percent of the Dutch market in replacement heavy tyres with none of its main competitors accounting for more than 8 percent. Dominance was also indicated by the wide range of tyres supplied by the company and its intensive advertising so that dealers ran serious commercial risks if they did not include Michelin tyres in their range of products.

It seems therefore that the authorities have been edging towards a structural definition of dominance. Structural factors certainly seem, in practice, to be the major considerations with other evidence playing a secondary role. However, the use of market-share data is acceptable only in so far as it is used as part of a filtering system for identifying *possible* areas of concern. A very high market share does not in itself indicate abuse, and indeed in the extreme case of contestable markets even a simple monopoly could be innocent of any anticompetitive performance. However, the theory of contestability is not very robust; small changes in the assumptions can drastically affect the predictions.[29] What the theory has succeeded in doing is to give added emphasis to entry conditions as a determinant of performance, though it is doubtful whether entry possibilities should be given as much weight as actual competition because they are often uncommonly difficult to assess. But whatever the weights to be attached to them, actual and potential competition plus other factors such as the rate of technological change and the rate of growth of markets, which influence entry possibilities, have to be examined before any association can be drawn between the existence of a dominant position and the existence of monopoly power.

---

[29] See, Vickers, J. and Yarrow, G. (1985), *Privatisation and the Natural Monopolies*, London: Public Policy Centre.
For an outline of the Contestable markets theory see Baumol, W. J. (1982), 'Contestable Markets: An Uprising in the Theory of Industry Structure', *American Economic Review*, **72**; and for a defence of the more traditional approach to the analysis of markets see Shepherd, W. (1984), 'Contestability vs Competition', *American Economic Review*, **74**.

It is also interesting to note the emphasis which has been placed on the gap between the market share of the leading firm and that of its nearest rival. This would seem to imply that the Court is of the view that dominant firm market structures pose a greater threat to competition than tightly-knit oligopolistic ones where, say, two or three well-matched firms account for the bulk of sales. Although Article 86 covers oligopolistic as well as monopolistic abuses of dominance no case against an oligopoly has been heard under EC law.

### 4.3. The Concept of Abuse

The wording of Article 86 makes it clear that it is the abuse of a dominant position that is condemned and not the existence of such a position, so that an uncompetitive market structure, with a dominant firm, will be permitted so long as the firm doesn't take advantage of its position to behave uncompetitively. Some examples of abuse are given in Article 86 itself—including high prices, price discrimination, restrictions on output and tied sales. Not all these practices are unambiguously anticompetitive. Price discrimination may, for instance, improve static resource allocation and may serve to sharpen competition in the dynamic sense. However the latter possibility, it has to be admitted, is more likely when practised by small to medium-sized firms and on an unsystematic basis as these firms attempt to gain market share.

The Court's pronouncements in this area have served to cloud rather than clarify the issues. In its first general definition of the concept of abuse (in the *Hoffman-La Roche* case) the Court declared that:

> 'the concept of abuse is an objective concept relating to the behaviour of an undertaking in a dominant position which is such as to influence the structure of a market where, as a result of the very presence of the undertaking in question, the degree of competition is weakened and which, through recourse to methods differing from those which condition normal competition . . . has the effect of hindering the maintenance of the degree of competition still existing in the market or the growth of that competition.'[30]

---

[30] Commission of the European Communities, *Ninth Report on Competition Policy*, Brussels, 1979, p. 29.

Since the Court has ruled that *very* high market shares can themselves constitute evidence of dominance, the above definition of abuse would seem to indicate that firms having those market shares are automatically guilty of abuse by virtue of their 'very presence.' This is in line with the judgement of the Court in the *Continental Can* case when it declared that a dominant firm which strengthened its market position can be guilty of abusing its position whatever methods it used to do so: 'the strengthening of the position held by the enterprise can be an abuse and prohibited under Article 86 of the Treaty regardless of the method or means used to attain it' if the degree of dominance essentially affects competition. This dictum raises the possibility, in theory at least, that Article 86 could be used to prevent dominant firms from extending their dominance even through internal growth unaccompanied by any anticompetitive behaviour.

However, the Commission itself has not attempted to interpret Article 86 in this way, and there is no case of a firm simply being accused of dominance with no attempt to prove deliberate anticompetitive behaviour.

The reference in the Court's definition of abuse to the dominant firm having 'recourse to methods differing from those which condition normal competition' is also unhelpful. This could be interpreted as saying that any behaviour which could not occur in a competitive market is abusive and prohibited. But this would still leave unresolved the important point that the *same* methods may be viewed differently as to their effects on competition depending on the particular circumstances of the market and the size of the firm employing them. Thus, for instance, price discrimination may enhance or suppress competition and the Commission has accepted that circumstances may exist when even market-sharing may have procompetitive effects. Whatever the interpretation that can be placed on the judgements of the Court the Commission has in fact taken a pragmatic approach. In practice it is serious abuse rather than any abuse that has been condemned and forms of business conduct are examined against the background of the circumstances of each case rather than in abstract.

### 4.4. Remedies

The remedies applied to monopoly problems in the EC are conduct remedies which control aspects of firms' behaviour. The Commission can, subject to appeal, prevent a dominant firm from continuing to abuse its position, either informally or by issuing a formal decision, which has the status of law. It also has the power, again subject to appeal, to fine offenders up to a maximum of 10 percent of the value of turnover.

Although the powers available to the authorities appear adequate enough it is doubtful whether the policy has been particularly effective in deterring dominant-firm abuse. The application of Article 86 was slow to get off the mark, the first case being heard in 1971, and up to the end of 1987 only some 30 odd cases had been examined.

A policy of relying on conduct measures can always be criticised as tinkering with the situation and unlikely to achieve any lasting improvement. If the uncompetitive market structure which causes or permits abuse cannot be changed, (and there is no provision for the disolution of existing concentrations in EC law), then firms will have to be constantly watched to unearth abusive conduct and to ensure that behaviour which has been condemned does not re-emerge. True, much of the police work can be done by aggrieved third parties but even so the conduct approach is cumbersome and expensive on administrative resources. Most important of all, perhaps, the attack on certain business practices can lead to the adoption of other forms of behaviour having the same result. It is impossible to control effectively the behaviour of dominant firms without some form of structural remedy, and in the absence of powers to dissolve existing concentrations it is better to adopt measures that prevent the emergence of dominant positions than to attempt to control the conduct of firms after dominance has been attained.

## 5. MERGER POLICY

### 5.1. Merger Control to 1989

The absence of powers to control mergers has been a major gap in EC competition policy. Article 85 covers agreements and concerted practices between enterprises and Articles 86 deals with abuse of dominance, but merger policy does not fall neatly into either.

The Commission has however succeeded in convincing the European Court of Justice that both Articles can be interpreted in such a way as to make them applicable to at least some aspects of merger policy.

The relevance of Article 86 was the first to be established in the *Continental Can* case (1971). We have already seen that the abuse of a dominant position is clearly condemned. Some examples of such abuses are given in Article 86 itself: the imposition of inequitable prices, the restriction of output, discrimination and tying clauses. These practices correspond to a possible direct misuse of dominance which damages consumers' interests. But a broader conception of an abuse could take into account, not only the direct impact of a market conduct on market performance, but also the indirect effect, through the effect of market conduct on market structure. In this view, Article 86 could be used to attack business strategies which modify, in a more or less irreversible way, the conditions of supply and demand in such a way that in the end social welfare is harmed.

This is the line of reasoning followed by the Commission and the Court. In the Continental Can case, where Continental Can took control of the leading Benelux producer of metal cans through its wholly-owned Belgian subsidiary, the Court stated that since the concept of abuse is not defined more precisely in Article 86, the Community's objectives set out in the Treaty of Rome must be considered before the examples of abuse given in Article 86 itself. An abuse is present where an enterprise conducts itself in a way that is 'objectively' wrong in relation to the goals of the Treaty: the 'provision is not only aimed at practices which may cause damage to consumers directly, but also at those which are detrimental to them through their impact on an effective competition structure, such as is mentioned in Article 3(f) of the Treaty. Abuse may therefore occur if an undertaking in a dominant position strengthens its position in such a way that the degree of dominance reached substantially fetters competition, i.e. that only undertakings remain in the market whose behaviour depends on the dominant one.'

Thus the Commission considers that market structure as such has to be protected: a change in the supply structure which virtually eliminates alternative sources of supply to the consumer appears to be an abuse in itself. Although the Court annulled the Commission's decision in the Continental Can case—mainly because the relevant market was not adequately defined—it very clearly confirmed the soundness of its interpretation: Article 86 may be applied to mergers.

The relevance of Article 85 was established much later following the Court's ruling in the *Philip Morris* case which occurred in 1981. The tobacco group had acquired a 25 percent stake in Rothmans International and although the Commission attached conditions to the deal this did not satisfy other tobacco companies who brought an action against it so that the case went to the Court. In 1987 the Court ruled that the Commission had acted correctly in the case and most significantly that share transactions can amount to a restrictive agreement under Article 85 irrespective of whether those transactions constitute an abuse of dominance under Article 86.

Potentially, at least, these two judgements give the Commission substantial powers of intervention in a wide range of merger cases. It is certainly the case that, increasingly, companies have deemed it wise to have cross-border mergers scrutinised by the Commission for compatability with Article 86 before proceeding with a bid. In other cases, mergers are scrutinised following complaints from affected parties, and in these cases the Commission has been able on occasion to make use of 'interim measures'. These measures, which can be used in cases of urgency, are designed to avoid irreparable damage being caused to a complainant or to the public interest when preliminary investigations are being undertaken. In 1981, for instance, the Amicon Corporation complained that a proposed merger between a Swedish company (Fortia AB) and a British firm (Wright Scientific Ltd) would increase Fortia's share of the European market for chromatography columns from 73 percent to 82 percent. The Commission sent a copy of the complaint to Fortia and Wright warning them of possible interim measures that might be taken if the facts alleged by the complainant turned out to be true. The merger negotiations were dropped.[31] Again the bitterly-fought battle for Irish Distillers was frustrated by a complaint to the Commission under Article 85 and the Commission's decision to threaten interim measures. Companies may also deploy EC competition law in the national courts as Plessey did in its attempt to ward off the hostile takeover bid by GEC/Siemens.

When the Commission has intervened in merger cases it has on occasion demonstrated its procompetitive stance rather more effectively than the national authorities. One example of this is the British Airways takeover of British Caledonian. BA made a bid for its main

---

[31] Commission of the European Communities, *Eleventh Report on Competition Policy*, Brussels 1981, p. 73.

UK competitor BCal in July 1987, and the bid was referred to the Monopolies and Mergers Commission. The MMC report, published in November 1987, concluded that the merger was not against the public interest.[32] The MMC was clearly impressed by arguments advanced by BA as to the benefits of the merger. These included a strengthening of the position of the company as an international carrier, benefits of synergy, and saving BCal from liquidation. Critics of the MMC's findings point to the fact that BA was already the largest airliner in the world in terms of the number of international scheduled passengers carried; that the Civil Aviation Authority in its evidence was sceptical of the view that there were further benefits to be achieved from economies of scale; and that even if BCal was in a parlous financial position a takeover by BA was not the only solution—a takeover by a smaller competing airline was a clear alternative. The MMC recognised that the merger would reduce competition and that there would be a greater risk of predatory and anticompetitive behaviour from an enlarged BA. Nevertheless, given the undertakings offered by BA, it felt that these detriments were outweighed by the likely benefits. These included: a surrender of BCal licences to operate domestic routes and routes between Gatwick Airport and certain European destinations (but it retained its right to reapply for the licences surrendered); and a surrender of at least 5000 slots at Gatwick.

Immediately after the bid was cleared, BA and Scandinavian Airlines (SAS) joined battle to take over BCal. BA, supported by the government, won. However, following a complaint by Air Europe the case had also gone to the European Commission. The Commission was not so easily satisfied by BA's assurances and BA had to agree to reduce its slots at Gatwick by more than 10,000, to surrender additional routes between the UK and Western Europe and not to reapply for BCal's internal UK routes which it had surrendered.

However, though the European Commission on this occasion showed itself to be a stronger defender of the competitive faith than the UK authorities, it is also important to note that its contribution was to

---

[32] Monopolies and Mergers Commission, *British Airways plc and British Caledonian plc: a report on the proposed merger*, Cm. 247, HMSO, London 1987. For further discussion of this case see, Hay D. and Vickers J. (August 1988), 'The Reform of UK Competition Policy', *National Institute Economic Review*.

amend the terms on which the merger could be endorsed rather than to prevent it.

Even though on the face of it the Commission would seem to have quite sweeping powers for dealing with mergers under Articles 85 and 86 the position in practice is quite unsatisfactory. For instance, it would seem that Article 85 can apply to agreed mergers but not to hostile takeovers, for only in the former case would it be reasonable to argue that a restrictive agreement might be involved. Yet, in reality, the distinction between the two cases in terms of intent may be less than sharp. The application of Article 86 to mergers control also poses problems. Although the Commission's actions have on occasion prevented a merger from taking place it is not entirely clear what powers it has for stopping a merger as opposed to intervening after it has occurred. The procedures used to apply the existing powers have also been widely criticised as unstructured and ill-defined. The absence of clear rules and a timetable for dealing with merger cases has caused uncertainty in the business community and great expense to those caught up in the prolonged proceedings of the Commission.

This unsatisfactory state of affairs is reflected in the Commission's own reports on competition policy. In 1980 it remarked: 'Experience of recent cases has confirmed that, even if Article 86 *is difficult to apply on account of the conditions for its application*, it still enables the Commission to *monitor to a certain extent* large-scale mergers and, if necessary, to prevent them being carried through or to have changes made which are desirable from the point of view of competition.' (emphasis added.) And in its 1988 report the Commission once again emphasised that the *prior* control of mergers 'is crucial to the preservation of competitive structures.'

It is indeed a well-defined policy for the prior control of mergers that has been lacking in the Community's approach to competition policy. The first attempt to introduce a draft merger regulation occurred in 1973 but this did not receive support from member states, and subsequent attempts to revive it also came to nothing. Whatever powers the Commission may have under Articles 85 and 86 virtually nothing has been achieved in controlling the oligopolistic structures of EC markets. 'It is very serious', complained a frustrated Commission, 'if it is not possible to respond adequately to mergers that so increase the concentration of markets that there is little room for competition.'[33]

### 5.2. The 1989 Regulation for the Control of Mergers

The Commission finally succeeded in getting a merger regulation adopted in December 1989. This section first outlines the main features of the regulation and then goes on to offer some comments on it.[34]

The mergers covered by the regulations are those 'which have a Community dimension' as defined by a combination of turnover criteria and geographical distribution of turnover within the Community. A merger has a Community dimension where: (a) aggregate worldwide turnover of all the undertakings concerned exceeds ECU 5 bn., and (b), the aggregate Community-wide turnover of each of at least two of the undertakings concerned is more than ECU 250 million. Mergers where two-thirds or more of aggregate Community-wide turnover is within one and the same member state are excluded from the Regulation. The thresholds are to be reviewed within four years of the Regulation being adopted. There is a presumption that mergers between firms with 'small market shares' are unlikely to impede competition, a presumption which exists in particular when the merging firms have a combined market share in the EC as a whole or in any substantial part of it not exceeding 25 percent. ('Cartel mergers', i.e. share transactions whose purpose is to coordinate the behaviour of independent concerns are excluded from the Regulation. They will continue to be investigated under Article 85.) Member states shall not apply their national legislation on competition to mergers having a Community dimension unless expressly empowered to do so by the Commission, or unless the merger raises issues of legitimate national interest, such as public security, plurality of the media and prudential rules, which are not taken into consideration by the Regulation. (Articles 9 and 21.)

It is interesting to note that compared to proposals contained in an

---

[33] Commission of the European Communities, *Sixteenth Report on Competition Policy*, Brussels 1987, p. 16.

[34] For the full regulations see: Council of the European Communities, *Council Regulation (EEC) No/89, on the control of concentrations between undertakings*, Brussels, 21 December 1989.

earlier draft regulation,[35] the Regulation finally adopted represents, in some respects, a weakening of the Commission's control over mergers. In that earlier draft the Commission had proposed thresholds of ECU 1 bn. for aggregate worldwide turnover; ECU 100 million for Community-wide turnover of each of at least two undertakings, with mergers having three-quarters of Community-wide turnover in one member state being excluded. A number of member states, particularly those such as the UK and West Germany with well developed merger policies of their own, strenuously resisted these earlier proposals, because it would have resulted in a substantial dilution of their national controls. The very big change in the thresholds was a price the Commission had to pay in order to achieve its objective of a Community-wide merger policy, but it must be added that these thresholds could be reduced at a later date by a majority vote of the Member States.

Of the mergers that fall within the scope of the Regulation, those which create or strengthen a dominant position, as a result of which effective competition would be significantly impeded in the common market or in a substantial part of it, are prohibited. In making its appraisal the Commission will take into account a wide range of considerations including:

market structure,
actual and potential competition from firms located both within and outside the Community,
the market position and economic and financial power of the firms,
the opportunities available to suppliers and users,
access to supplies and markets,
barriers to entry.
supply and demand trends for the relevant goods and services,
the interests of customers,
the development of technical and economic progress provided that it is to the consumer's advantage and does not form an obstacle to competition.

The regulation also details matters of procedure including the time

---

[35] For details of this earlier draft regulation see Commission of the European Communities, *Amended proposal for a Council Regulation (EEC) on the control of concentrations between undertakings*, COM (88) Final-Revised Version, Brussels, 19 December, 1988.

which the Commission is given to deal with cases. It has one month from the date of notification to decide whether or not to open proceedings. If it does so decide it has a further four months to consider the evidence and either block the merger or exempt it. This time limit may be extended if, for instance, the firms under investigation have been obstructive or have supplied incomplete or inaccurate information.

Given that the Commission has been attempting to introduce prior merger control since 1973 it is perhaps surprising (to all except those who are aware of the constraints imposed in any political compromise) that the draft Regulation still contains ambiguities and weaknesses.

The use of firm size, for instance, as the criterion for determining eligibility for notification is clearly suspect both on economic and legal grounds. The most obvious criticism is that the size of the firms has to be viewed within the context of the relevant market. Inevitably, a good many mergers will fall outside the turnover thresholds yet will result in firms with substantial market power because they operate in product markets in which total EC turnover is relatively low. The legal ambiguity in all this is that the European Court has (see Hoffman-La Roche case in Section 4) denied the relevance of a company's overall size in establishing abuse.

The same point arises in the criteria listed for appraising mergers. In so far as financial power is related to size past experience suggests that this will not convince the Court, nor will any attempt to relate economic and financial power to size alone. At best the Court is likely, following the Hoffman-La Roche and United Brands cases, to view a firm's performance as of secondary evidential value. Apart from this, the criteria are sufficiently wide to capture the most important factors which are relevant in establishing dominance—in particular market structure, including vertical relationships; entry barriers; international competition and market trends in the supply and demand for the goods or services concerned. However it is also true that criteria such as 'market position', 'economic and financial power of the firms concerned' and 'structure of all the markets concerned', are ambiguous indicators without clear economic content. Market position, for example, can only be determined by reference to a number of structural features. The overall impression is that the list of criteria is unsystematic and will have to be more sharply defined by case law.

An important feature of the regulation is that although considerations of efficiency enter into the appraisal of a merger there is no 'efficiency defence'. Here again it is interesting to note a change in emphasis from the earlier draft regulation proposed by the Commission. That proposal contained an explicit efficiency defence, so that a merger which created or strengthened a position of dominance and which significantly reduced competition might still be allowed if the competitive detriment was outweighed by the contribution the merger made to attaining other objectives such as 'improving the competitive structure within the common market', or 'promoting technical or economic progress'. In judging the merits of the efficiency defence, the Commission would have had to take account of the competitiveness of the sectors concerned with regard to international competition and the interests of consumers, and would not have been able to accept an efficiency defence if the merger afforded the firms concerned the possibility of eliminating competition in respect of a substantial part of the goods or services concerned. This earlier proposal, however, met with strong opposition from some member states. It was felt that the efficiency defence was widely drawn and that the Commission might be tempted into using it for purposes of industrial strategy—such as creating 'European winners' to compete with the Japanese and American giants—and that this might be done at the expense of a significant weakening in competition policy. The critics of this earlier proposal argued that the Commission should use only one criterion for judging mergers—the effect on competition, leaving decisions on wider issues to politically more accountable bodies.

The wording of the adopted regulation certainly suggests that the emphasis is unambiguously on preserving and developing effective competition. A merger which threatens competition can not be permitted because of overriding efficiency gains. The 'development of technical and economic progress' is an argument to support a merger only 'provided that it is to the consumers' advantage, and does not form an obstacle to competition', i.e. the goal seems to be to maximise consumers' surplus rather than the sum of consumers and producers' surplus. But if this is so there seems little point in including the efficiency criterion at all. All the Commission needs to establish is whether a merger is likely to form an obstacle to competition, and this it will have to do by reference to all the other criteria. The regulation has all the hallmarks of an inevitable compromise. The Commission started off with a proposal for an efficiency defence; this was defeated;

but efficiency becomes one element in the overall appraisal. If there is no efficiency defence, however, the inclusion of an efficiency element in the appraisal is superfluous. It counts only when there is no conflict with competition. Since, when there is a conflict, competition must always win the competition criteria are the only ones needed.

In practice, the position may well be different. The difference between the earlier draft regulation and the adopted version may in fact be more apparent than real, because the criteria for assessing the competition effects of mergers are inevitably vague. Given the difficulties that will commonly be encountered in evaluating the strength and likely trends in competition both within and from outside the common market, efficiency criteria may well in practice be more important than the precise wording of the regulation suggests. Doubts concerning the competitive impact together with a strongly argued case of technical and economic progress may result in several mergers being approved mainly on grounds of dynamic efficiency gains and the need to meet competition from large firms outside the Community—a situation little different to that which would have prevailed with an explicit efficiency defence.

On matters of procedure it seems clear that the Commission intends to adopt a neutral approach. This is suggested in the preamble to the regulation which states that the greater unification of the EC market will result in major and beneficial restructuring via merger, and that this will strengthen the competitiveness of European industry. There is little foundation in theory or in empirical evidence, however, to support this optimistic view of the effects of mergers in general. Experience of UK merger policy suggests for instance that claims of efficiency gains are easily asserted but far more difficult to demonstrate. Furthermore, a wide range of empirical work on the effects of mergers does not give cause for much optimism. There is a real danger that the much publicised 1992 (which in terms of the extent to which it will mark a change in underlying economic realities has been overplayed) will be seized upon by many firms as the ideal excuse for amalgamation with no serious attempt by the Commission to justify them. Since the proposed regulation will capture only the very largest mergers the least that can be expected is that the onus of proof should be on merging firms to demonstrate substantial benefit which could not be achieved in other ways. They, after all, should have the detailed information necessary to demonstrate such benefit. If they cannot do so, and given the absence of any effective policy for unscrambling a merger once it has occurred, the benefit of any doubt should be given to maintaining competitive market structures.[36]

The merger regulation will greatly enlarge the powers of the Commission. This has been described as both unnecessary and undesirable. Unnecessary because satisfactory procedures for dealing with mergers are already in place or evolving in member states, and undesirable because it will place undue power in the hands of a body which is ill equipped to deal with existing responsibilities let alone new ones, and highly prone to political lobbying. The basis for such accusations is not difficult to find. There are considerable doubts about the Commission's ability to deal effectively and fairly with even the relatively small number of mergers which will fall within the scope of the new regulation. It is estimated that with a ECU 5 bn. threshold fifty to sixty mergers a year will be caught in the Commission's net. This compares to fewer than twenty decisions a year made under Articles 85 and 86. Some merger cases will be cleared within four weeks. Will it be possible in this time for the Commission to give full consideration to the views of member states and to third parties affected by the merger? This is why some Member States are anxious that mergers which are given clearance, with no formal decision being taken, should fall under national jurisdiction. There is in fact provision (Article 9) in the regulation for the Commission to refer certain notifiable mergers to the competition authorities of a Member state with a view to the application of that state's national competition law. The circumstances when this applies is where a Member State forms a distinct market within the Community. In the event of a conflict between the Commission and a Member State the latter may appeal to the Court of Justice for the purpose of applying its own legislation.

As to those mergers that are fully investigated, the Commission in its investigations will have to rely heavily on cooperation from national authorities. It will have to convince Member States that it will be able to deal with these cases not only within the prescribed period but also in a way which attracts the confidence of all interested parties by giving them a fair hearing.

---

[36] See, for instance, George K. D. (1989) 'Do We Need a Merger Policy?' in Fairburn J. A. and Kay J. A. (eds) *Mergers and Merger Policy*, Oxford University Press, 1989.

The Commission is also open to criticism because the powers of prosecutor, judge and jury are too closely intertwined in its hands. It will have a great deal of discretionary power both in determining dominance and in assessing competitive effects, and this is bound to give even more scope for political lobbying, which in any case exists. However, there is an important safeguard in the provision for the Commission's decisions to be appealed to the Court of Justice.

To be fair, these problems of conducting a merger policy are not peculiar to the Commission. Political lobbying is a frequent-enough occurrence in member states and the task of assessing the pros and cons of mergers inevitably mean that considerable discretion has to be used by the responsible authority. It is for this reason of course that it is important that merger cases are seen to be handled fairly and openly and why there is a strong case for the final decision to be taken by a body other than the investigating authority, as is the case for instance in the UK and West Germany.

### 5.3. Concluding Comments

There is a good case for an EC merger policy but the present regulation, although being an important achievement, does not provide an entirely satisfactory basis for it. Some of the problems with it have been pointed out above. These problems could be solved, in part at least, by the adoption of guidelines for the investigation of individual merger cases. Experience in dealing with merger investigations should also result in improvements. Another point that must again be underlined is the danger of accepting too easily the view that an increase in merger activity is a central way of restructuring European industry and making it more competitive in world markets. This may be so in certain cases, but as a generalisation it does not stand up to careful scrutiny. Empirical evidence shows no clear positive association between size and efficiency nor between the latter and merger activity. And any presumption that international competition can be relied upon to ensure workably competitive markets is also misplaced. Within the Community much emphasis is now being placed on the great liberalisation of markets following 1992. Some of the changes which it is said will occur have almost certainly been exaggerated, particularly the extent to which yet further economies will be achieved by longer production run and larger firms. The efficiency benefits of a wider and freer market come mainly in the form of greater competition between firms. It is necessary to avoid the risk that the pronouncements and policies of the Commission attenuate those benefits. An earlier survey of EC competition policy suggested that 'the policy conclusion to be drawn for Europe appears to be the urgency of passing and rigorously enforcing laws to control industrial structure before the benevolent deconcentrating effects of economic integration is totally eroded'.[37] That concern still exists.

---

[37] George, K. D. and Joll C. (1975), *Competition Policy in the UK and EEC*, Cambridge University Press, p. 213.

# Conclusions for Competition Policy

WILLIAM S. COMANOR

*University of California, Santa Barbara, with the assistance of* Ken George, *University College of Swansea,* F. Jenny, *E.S.S.E.C., and* Leonard Waverman, *University of Toronto*

There is often great ambivalence in the adoption and enforcement of a competition policy. Despite general acknowledgement that an effective competition policy promotes desired economic performance in a market economy, there is much doubt and some controversy as to what should be done. What criteria should be applied and what remedies should be enforced? Because there are no undisputed answers to such questions, policy judgments have differed widely. In this volume, we have explored these differences and reviewed how the principal economies of Europe and North America have reached different answers.

There is little doubt but that monopolistic structures and practices generally retard economic performance. However, there is also little doubt that these same structures and practices can be innocuous or even lead to improved performance in certain situations. For every result, there is also an exception. An essential question is therefore whether a policy can be designed that can discriminate between these various circumstances. The question here is whether a policy can be crafted so that it is applied only when it enhances economic performance. Alternatively, if general rules must be applied, because it is difficult to discriminate among the underlying situations, then the relevant issue is whether rules can be fashioned such that performance is enhanced more frequently than diminished.

A second source of ambivalence with competition policy follows from different judgments of the prevalence of monopolistic structures and practices. If these forms are commonly found, then there is a larger role for competition policy to play. On the other hand, if monopolistic structures and practices are rare, and effective competition is the norm,

## CONCLUSIONS FOR COMPETITION POLICY

there is less need for a competition policy; and policy actions are more likely to have undesired consequences.

If monopolistic elements are more frequent in some economies than others, then policy actions should differ among countries. A relevant matter therefore is whether smaller economies, such as Britain and Canada, have monopolistic structures more commonly than larger ones like the United States. Another question is whether certain economies are open to trade, which describes whether or not domestic markets are protected from foreign competition.

A related issue concerns not the prevalence of monopolistic forms but rather their rate of decay. Even if monopolistic structures and conduct are commonly found, their economic consequences could dissipate rapidly. In this case, there would be little need for an active competition policy.

Although frequently stated in terms of specific issues, such general concerns are the source of prolonged debate. As a result, there are different answers given to the question of how stringently competition policy should be enforced. Ambivalence on this matter stems from real differences of opinion about both the economy on which the policy must operate and the prospect that effective actions can ever be designed.

It is therefore hardly surprising that different countries have different rules and institutions to carry out a competition policy. And also that countries have pursued very different policies at different periods of time. The quest for an optimal policy design is ongoing, and there is no suggestion from our studies of enforcement practices in Europe and North America that there is only one preferred approach.

The object of this volume has been to compare different policy regimes. All of the countries considered here are market economies, and we observe monopolistic forms in all of them. Yet, there are substantial differences in their policies designed to deal with this problem. In large measure, these differences stem from varying judgments over the prospect for designing an effective policy.

With this ambivalence, it is not surprising that the enforcement of competition policy in each country has waxed and waned over time. Still, it is ironic that just as the pace of enforcement has accelerated in Europe under the aegis of the European Community, there has been a notable decline of the effectivness of antitrust enforcement in North America, principally in the United States. Although the US was long the most avowed proponent of a vigorous competition policy, that may no longer be the case. Europe could well pursue a more vigorous competition policy in the years ahead.

A striking feature of the recent acceleration of antitrust enforcement in Europe is its relation to the increasing economic integration of the continent. A general policy statement supporting competition was included in the Treaty of Rome and new policy directions have followed in its wake. This acceleration is parallel to that which occured during the 19th century in the United States, when an American national market was forged out of local and regional markets.

The expanding role of competition policy in the European Community has also affected national policy directions. As Frederic Jenny points out, the 1977 reforms in France were carried out under a Prime Minister who had previously been chairman of the Commission for the European Economic Community.

The growing emphasis on competition policy in the European Community suggests an interesting hypothesis between this policy and the extent of economic integration. Earlier, it was the United States that was the primary proponent for an effective competition policy. Other countries with smaller and more open economies felt less need for it. More recently, the European Community has followed in this direction, while at the same time, the United States has become a more open economy. And with the growing importance of international trade has come a decline in the pace of US antitrust enforcement.

## 1. RULES OR DISCRETIONARY AUTHORITY

Because the impact of a competition policy may differ according to the economic circumstances in which it is applied, there is considerable temptation to create an expert quasi-judicial body with wide discretionary powers to evaluate alternate policy actions in the context of specific markets. In this setting, competition policy could be made to fit particular industrial circumstances. Various countries have taken this road although it can create problems of conflicting jurisdiction between these bodies and the normal judicial system.

In the United States, the allure of an expert body to deal with these issues was the primary rationale behind the creation of the Federal Trade Commission in 1914. There would be a commission who would gain extensive expertise over such matters and which had wide authority to reject 'unfair methods of competition.' Similar authority was vested in the British Monopolies Commission, the Canadian Restrictive Trade Practices Commission and the French Conseil de la Concurrence.

The British Monopolies and Restrictive Practices Commission was created in 1948. It had the power to investigate monopolies and cartels and report on whether these situations might 'be expected to operate against the public interest.' However, the public interest was not defined and the word 'competition' did not appear.

Despite the rationale, or perhaps because of it, the effectiveness of these agencies was quite different than expected. Early on in the history of the US Federal Trade Commission, its powers were constrained by judicial review so that they were hardly wider or more flexible than those embodied in the other antitrust statutes. In Britain, Commission powers were entirely those of investigation and recommendation, with final decisions left to politicians. A result of that structure is that the Commission did not focus attention on competition issues but rather included such concerns as employment prospects and regional effects in making its recommendations.

In Canada, the Restrictive Trade Practices Commission was originally a research body, although it acquired adjudicative powers in 1976. In 1986, it was renamed the Competition Tribunal and became the civil court of record in this area. An original problem in creating a quasi-judicial tribunal in Canada was the requirement of a division of powers in the Canadian constitution.

In France up to 1986, the Commission de la Concurrence was not constrained by judicial review as it was merely an advisory body. But this changed in 1986 as the successor body acquired enforcement responsibilities.

An interesting question is why discretionary authorities have been less vigorous in the enforcement of competition policies. A possible answer is that individual situations are often unclear or ambiguous so that most government officials are unwilling to exercise clear authority. This is particularly true when the statutory guidelines are as general as a ban on 'unfair methods of competition.' This problem is particularly evident in the United Kingdom where the 'public interest' is so widely defined that there is no limit on the matters which the Commission can consider. It is not enough to create a body or designate a policy-maker, and then provide him or it with broad powers. What is also required are specific guidelines on which to rely. However, legislative guidelines would require the legislatures to articulate explicitly the goals of a competition policy; and this is rarely done.

When competition policy is enforced through pre-determined rules, it has been far more effective. Not only are enforcement officials more inclined to pursue established policy objectives, but equally important, so are the private firms which are the subject of the policy. To avoid specified penalties, the firms affected will generally alter their conduct so that policy objectives can be achieved without explicit government actions. This type of impact on private behavior is far less likely when competition policy is administered through the discretionary powers of government officials. Policy decisions cannot be predicted so there is greater temptation to pursue short-term interests. Furthermore, the likely penalties are problematical.

Even where there are specific rules to be enforced, much depends on how precisely they are defined. To be sure, the statutory provisions that underlie competition policy, whether in the Treaty of Rome or the US Sherman Act, are hardly explicit. What are 'restraints of trade' and what conduct comprises 'monopolizing behavior?' For the most part, the relevant policy rules are set down not in statutory provisions but rather in judicial decisions. But they are definite rules just the same, and what is relevant for an effective competition policy is what they require rather than how they are established. Moreover, an investigative mechanism is required as well as a set of penalties which impose substantial costs on those who break the rules.

From the earlier chapters in this volume, we can compare the particular rules as they pertain to specific issues of competition policy. And we can determine how precise or abstract these rules are in varying circumstances. What is apparent is that competition policy is more strongly enforced when these rules are more precisely set down, and less so when the relevant strictures are unclear so that more authority rests with discretionary bodies. This being so, a critical issue for competition policy-making is the definition of appropriate rules.

## 2. RULE-MAKING FOR COMPETITION POLICY

In the United States, the essential rules for competition policy are largely contained in Supreme Court decisions. These opinions add considerable substance and detail to the original statutory language, and comprise a specific set of rules which set the bounds for competition policy in the United States. More recently, similar developments have taken place in the European Community where decisions of the European Court of Justice have added substance to Articles 85 and 86 of the Treaty of Rome. In Britain, however, policy development has lagged, perhaps because it depends less on case law and judicial decisions and more on occasional government reviews of policy standards. In France, although the Paris Court of Appeal now reviews the decisions of the Conseil de la Concurrence, the latter body still has a major influence on the interpretation of the law.

To be sure, the rules contained in legal decisions are not necessarily alike. For example, judicial decisions in the United States have emphasized the dichotomy between *per se* violations and that conduct adjudicated under the 'rule of reason.' Similar distinctions exist elsewhere. For example, the British approach towards resale price maintenance is nearly one of *per se* illegality which policy standards towards other forms of restrictions are more equivocal. In France, on the other hand, horizontal price agreements are considered *per se* offenses while vertical agreements are evaluated under the rule of reason.

In Canada, the legislative language is also imprecise so that the courts must serve as the arbiter of competition policy. Occasionally, the legislature has rewritten sections of the law dealing with specific concerns. Moreover, few *per se* prohibitions exist. Instead, the rule of reason holds for the vast majority of potential offenses.

These distinctions have critical importance for the vigor by which competition policy is enforced. Thus, Professor Liebler writes that in the United States plaintiffs win most *per se* cases while defendants always prevail under the rule of reason.[1]

This generalization does not hold for Canada, where a more critical issue is the burden of proof, whether placed on the plaintiff or defendant, and the evidence that the court requires to demonstrate a particular policy infraction. For example, few group conspiracies are written down. As a result, courts have to determine from

---

[1] Wesley J. Liebler, 'Resale Price Maintenance and Consumer Welfare', *Business Electronics Corp. v. Sharp Electronics Corp.' UCLA Law Review*, forthcoming, p. 5.

circumstantial evidence whether or not an offence exists. Who has the burden of proof to demonstrate either an anticompetitive effect or the absence of such an effect has a critical influence on the final outcome.

## 3. PROSPECTS FOR AN EFFECTIVE COMPETITION POLICY

As the chapter by Janusz Ordover points out, there are unresolved issues which must be accepted to support or reject an active competition policy. While the evidence on some of these issues remains uncertain, there is wide acceptance of the viewpoint that more competitive markets promote improved economic performance, although not necessarily in every individual case. This judgment was made in North America at the latter stages of the 19th century and has been supported in Europe more recently. It rests not only on prospective effects on efficiency and dynamic advance or progressiveness, but also on matters of equity and the distributions of income and wealth. Monopoly prices impose a tax on consumers, but one which is not used to support public purposes, and improved equity is surely a dimension of economic performance. For all of these reasons, there is general agreement on the beneficence of competition and the need for an effective and rational competition policy.

Where this concensus breaks down is just how this policy should be fashioned, and what actions should be taken to promote improved performance with the minimum incidence of deleterious side-effects. As reported above, there are various answers to this question, and we should not expect that a single approach would be adopted. Still, the growing importance placed upon an effective competition policy in advanced market economies demonstrates the widespread view that enhanced competition promotes economic performance and that the development of an effective competition policy is a significant step in this direction.

# Index

Aaronovitch, S. 127
Abuses of dominant position 177–185
  concept 231–233
  definition of relevant markets 179–183
Accounting rate of return 35
Advance Ruling Certificate (ARC) 97
Advisory Opinion (AP) 97
Addy, G.N. 101
AEG-Telefunken case 226
Aetna Insurance Company case 86
Agreement
  anticompetitive 166
  cartel, prohibition of 191–193
  cooperative and state aids 214–215
  horizontal 169–174
    exchanges of information 174
    explicit 168–171
    tacit, and parallel behaviour 171–173
    proof of 81–84
  selective-distribution 176, 225–227
  vertical restrictive 174–178
Air Europe 236
Alcoa case 53, 54
American Telephone and Telegraph Company (AT&T) case 46
American Tobacco Company 53
Amicon Corporation 235
Animal waste, investigation 121
Anti-combines law 77
Anticompetitive agreement 166
Anticompetitive practices 182–184
Antimonopoly law, Canadian 88
Antitrust
  development of Canadian law 73
  history 75
  law, American, differences in basic philosophy 190–191
  market 35
  policy 8
    United States: issues and instructions 43–72
Act 1890 43
Enforced Division 45, 60
  remedy and liability in 55
Statutes 1953, in France 148–150
Apples, demand for 36
Arbitrage 37
Architects, decision on 168
Areeda, P. 57, 65
Ashland Chemical France, and Cabot Corporation 186
Atlantic Sugar case 83, 84, 86

Bain-Syloa-Modigliani (BSM) limit pricing model 21, 22
Baker, D.I. 64
Baker, J. 37
Baldwin, W.L. 11, 24
Balfour Committee Report on Industry and Trade 104
Barre reform 1977 150
Barton, D.M. 59
Bassett, L.R. 100
Baumol, W.F. 7, 11, 16, 20, 24, 25, 26, 29
Baxter, W.F. 60, 63
B.C. Sugar case 91, 94, 95
Beer supply case 125, 137
Behavioural aspects, of legislation 5
Benelux firms 221
  metal can producers 234
Berkey Photo decision 54
Bicycles investigation 143
Bladen, V.W. 75
Blair, R.D. 56
Blumenthal, W. 64
Bombardier snowmobiles 101
Bork, R. 4, 6, 8, 9, 70
Boston Consulting Group 201
Breakfast cereals 124
  monopolization case 55

Brennan, T.J. 33
Bresnahan, T.F. 32, 37
Brewers 138
Brick manufacturers, in France 172
British Airways, and British Caledonian 235, 236
British oxygen case 120
Brown Shoe decision 65, 66
Brozen, Y. 203
Bureau of Broadcast Measurement (BBM) 102
Business practices, restraints on 153-156
Buyers and sellers, mergers between 65-67, 160
Buying power, abuse of 159-160

Canada 253
　Competition law: 100 years 73-103
　Constitution Act 87
　Director of Investigation and Research 77
　Economic Council 2, 77
　history of law in 78-80
　　Department of Justice 78
　　Restrictive Trade Practices Commissioner (RTPC) 80, 249
　Montreal, City 82
　Supreme Court 85
Canadian antimonopoly law 88
Canadian Breweries case 90, 94, 95
Canadian Electrical Distributors Association 82
Canadian General Electric Company case 82
Car Parts report 140
Carbon black industry 187
Carlton, D.W. 52
Cartel agreements, prohibition of 191-193
Cat and dog foods supply case 121
Catherwood, H.F.R. 107
Celler-Kefauver Amendment, to Clayton Act 190
Cement Makers Federation (CMF) 111
Chicago School model 20, 21, 22, 202, 203
　/Demsetz paradigm 21, 31
Chirac administration 160, 161
Cigarette
　filter rods case 120-121
　paper manufacturers 173
Clayton Act (1914) 43, 44, 58, 192, 196, 197

Celler-Kefauver Amendment 190
Clin Midi 171
Colgate Palmolive, and Henkel-France 185
Collective discrimination 105
Collusive activities 81-88
　definition of undue 84-87
　exemptions 87-88
　proof of agreement 81-84
Collusive behaviour
　horizontal 109-118
　　cooperative agreements and state aids 214-215
　　effectiveness of policy 112-114
　　1988 Green Paper 115-117
　　industry restructing 215-216
　　legislation 109-112
　　R & D and new industries 216-219
　　small and medium-sized enterprises 219-220
　prohibition of 48-53
Colour film report 120, 123, 124
Comanor, W.S. 14, 43-72, 246-252
Combines Investigation Act 85, 87, 88
Commission de la Concurrence 151, 152, 153, 163, 167, 172, 176, 181, 182, 184, 185, 187
Commissions d'Urbanisme Commercial (CUC) 154, 155
Competition
　Act 76, 80, 81, 88, 93, 96, 98, 108, 118, 126, 136, 143
　Canadian law: 100 years 73-103
　definition of undue 84-87
　French 146-190
　　law 158
　　recent changes 187, 188
　and monopoly, normative economics 9-25
　policy
　　Bureau 94
Conseil de la Concurrence 5, 156, 157, 163, 165, 166, 168, 171, 173, 175, 176, 177, 179, 185, 186, 188, 189, 249
Constitution Act (Canada) 87
Consumer welfare, through allocative and productive efficiency 4
Continental Can case 209, 228, 232, 234
Contraceptive sheaths case 120
Cooper, T.C. 52
Cooperation between Enterprises, Notice on 218

# INDEX

Cooperative agreements and state aids 214-215
Cosmetics, distribution by pharmacists 175, 176
Cournot game 27
Cowling, K. 131
CPUs (central processing units) 15
Cross-product effects 29-30
Curry, B. 127

d'Aspremont, Q. 217
'Deep pocket doctrine' 201
de Jong, H.W. 212, 213, 214, 221
Demsetz, H. 20, 59, 203
  paradigm 21, 31
Discounts 227-228
Distillers and Guinness 129, 224, 235
Distribution
  exclusive 223-224
  selective, and franchising 225, 227
Dominance, existence of 229-231
Dominant enterprises, control of 193-195
Dominant positions, abuses 177-185
  criteria for 180-182
  remedies for 184
Dr Miles case 68
DTI 'Department for Enterprise' 130
Dunlop, B. 81, 87, 100
Duolite International, and Rhom et Hass France 185, 187
Dynamic efficiency and market power 22-25

Eastman, H. 75
Economic
  Council of Canada 1, 77
  Development Committees (EDCs)
  foundations of competition policy 7-42
  freedom of market competitors 3
  policy, French, postwar 147-148
  power, diffusion of 3
Economics, normative of competition and monopoly 9-25
Eddy Match case 89, 90
Edwards, C.D. 2, 201
'Efficiency defense' 5
Electric Reduction Company of Canada case 89
Elzinga, K. 59

Enterprise
  DTI 130
  SMEs 209, 219, 220
'Eurocrats, faceless' 151
European Court of Justice 174, 233, 251
European Economic Community
  competition policy in 206-245
  objectives 207-209
  first report 207
  seventeenth report 208
  horizontal collusive behaviour 209-222
  merger policy 233, 245
  monopoly and the abuse of market power 228-233
  progressive involvement of France in 147
  vertical arrangements 223-228
Exclusion, principle of 89
Exclusive-dealing arrangements 71-72, 139
Exclusive distribution 223, 224
Exclusive purchasing 225
Exemptions 87-88
Explicit agreements 168-171

Fair Trading Act (1973) 108, 18, 128, 136
  Office 108
Fairburn, J.A. 135
Farrell, J. 34
Fast-food market 140
Federal Trade Commission Act (1917) 43, 44, 59, 248
Fenton, K.M. 70
Ferruzi, and Saint Louis 185
Film processing case 119
Fine Papers case 86
Firth, M. 132
Fisher, A. 70
Fisher, F. 17, 35, 39
Fixed quality, assumption 16
Florists, in France 169
  Interflora 182
Fortia AB, and Wright Scientific 235
Fox, E.M. 8, 9, 48
France
  Administrative Supreme Court 163
  Competition policy 146-190
    emergence of law 146-162
    enforcement of law 162-187
    postwar economic policy 147-148

Conseil de la Concurrence 5, 156, 157, 163, 164, 165, 166, 173
  florists 167
  pharmacists 168, 171-172
Franchising 140-141
  and selective distribution 225-227
Franks, J.R. 132
French, H.E. 57, 72
Free trade policy 2
Fringe firms 27
Frozen foodstuffs case 139
Fudenberg, D. 17
Full-line forcing and tie-in-sales report 137

Gaskins, D. 18, 22
Gellhorn, E. 70
George, K. 92, 93, 104-145, 206-245, 246-252
Geroski, P. 19
Gibbens, R.I. 68
Gilbert, R.J. 19, 22, 31
Goldman, C.S. 94
Gosse, R. 73, 96
Gotthold, J. 203
Greenwood, J. 35
Grinnell decision 54
Grover, W. 98
Grundig-Consten test case 223
Guest, Keen and Nettlefold (GKN) and Sachs 199-202
Guinness and Distillers 129, 234-235

Hannah, L. 127
Hard-core practices 116
Hart-Scott-Rodino Act (1976) 58, 96, 201
Harris, R.S. 132
Hay, D. 10, 135, 144, 236
Henkel 185
Herfindahl-Hirschman Index 21, 28, 63, 197, 198
Hilferding (?) 3
Hoffman-La Roche case 229, 231, 240
Hofstadter, R. 45
Holt, C.A. 52
Hoppmann, E. 202, 203
Horizontal agreements 168-173
  explicit 168-171
  tacit, and parallel behaviour 171-173
  exchanges of information 173
Horizontal collusive behaviour 209-224

Horizontal mergers 197, 200
  comparison between German and American 197
Household detergents case 119, 121
Hughes, A. 113, 132

Ice cream and water ices case 139
Imperial Oil Consent Order 98
Indices of marker power 25-35
  Lerner index 25-30
  market share 31-34
  profit 34
Indirect Reprographic Equipment case 121
Individual practices 161
Industrial Reorganisation Corporation (IRC) 106
Industry restructuring 215-216
  and R & D 216
Interflora 181
Intertemporal effects on cost and demand 30
Investment Canada Act 94

Jacquemin, A. 1-6, 19, 92, 93, 206-245
Jenny, F. 146-190, 246-252
John Lewis retail group 227
Joll, C. 212, 224, 229

Kaiser, G. 73, 78, 87, 89, 95, 96, 101
Kantzenbach, E. 189-205
Kaplow, L. 25, 36
Kasserman, D.L. 56
Katz, M. 217
Kay, J.A. 35, 127, 135, 143
Kaysen, C. 52
KC Irving case 91
Kellogg's 21, 124
Kennedy bill 201
Khemani, R.S. 97, 99
Kirzner, I. 203
Kodak 123, 124
Koenker, R.W. 17
Krattenmaker, T.G. 49, 56
Kumar, K.S. 131, 132

Lande, R.H. 8, 9, 48
Landes, W.M. 13, 25
  LP approach 27, 33, 36, 56
Latte Farms 32, 33

# INDEX

Law
  against Restraints of Competition
    (West Germany) 189, 190, 196
  Canadian competition: 100
    years 73-103
    history 78-80
    administration 79-80
  competition, French,
    emergence 146-162
  enforcement of in France 163-188
  merger
    prior to 1986 94-96
    since 1986 96-97
    problems 98
Learned Hand, Judge 53, 56
Legislation
  behavioural aspects of 5
  horizontal collusive
    behaviour 109-112
  market structure 5
  monopoly 118-126
  performance 5
Le Quesne, Sir G. 133, 134
Lerner, A. 55
-Lange decentralized process 1
Index 12, 13, 25-30, 34, 35, 56
Librium and Valium, profitability of 120
Liebler, W.J. 251
Littlechild, S.C. 131
Loi Royer 154

MacDonald, B. 92
Marginal cost pricing 17
Maricopa decision 50
Market
  abuse and monopoly 228-233
    defining 228
    dominance 231-232
    concept 230
    existence 232
  definition 35-37, 178-180
  economics, competition policy 1-6
  liberalization of after 1992 244
  milk 32, 33
  perfectly contestable (PCM) 20
  specific power, indices 23, 25-35
  cross-product effects 29-30
  entry 28-29
Markham, J. 201
Mathewson, G.F. 100
Matsushita decision 57
McGee, J.S. 57

McGowan, J. 35
McGuire Act (1952) 68
McKee, J.S. 100
Meeks, G. 132
Mergers 93-102, 126-136
  between buyers and sellers 65-67
  between rivals 58-65
  control 158-159, 185-187, 198-199
  horizontal 197-200
  law
    prior to 1986 94-96
    since 1986 96-97
  non-horizontal 200-202
  policy 127-130, 135-247
    control up to 1989 233-238
    proposals for reform 133-135
  Regulations 1989 238-244
  UK policy 242
Mestmacker, E. 3
Metal Box Company 121
Metro II 174
Michelin Nederland case 227-230
Milk market 32, 33
Minnesota Mining and Manufacturing
  (3M), and Spontex 185, 186,
    188
Monopolies
  Commission 106, 108, 125, 129, 133,
    134, 138, 144, 236-249
  assessment 120-121
  effectiveness of policy 124-126
  German 199-200
  remedies 123-124
  and Mergers Act (1965) 106, 127, 128
Monopolistic structures and
  practices 246
'Monopolization' 53, 252
  cereals case 55
Monopoly 88-93
  and abuse of market power 228-233
  remedies 232-233
  definition 119-120
  effectiveness of policy in 124-126
  normative economics of 9-25
  static deadweight cost 12-18
  dynamics of deadweight cost 18-22
  power exercise and abuse 53-58,
    118-126
  provisions under Competition Act
    (1986) 92
Möschel, W. 203
Mueller, D. 19

Narver, J.C. 201
National Economic Development
    Council (NEDC) 106, 113
Neo-Austrian School 203
Nestle, and Rowntree Mackintosh 186
Newspapers, advertising rates 183
Nielsen-Jones, I. 73, 87, 89, 95, 96, 101
Non-horizontal mergers 200–202
Normative economics of competition and monopoly 9–25
    dynamic efficiency and power 22–25
    dynamics of deadweight cost of monopoly 18–22
    static deadweight cost of monopoly 12–18
Nova Scotia Board of Insurance Underwriters 86
Norzick, R. 10

OECD 96
Office of Fair Trading (OFT) 108, 128, 129, 134, 135
    Director 118
Okun, A.M. 227
Oleofina 213
Oligopoly 27–28
Ordover, J. 7–42, 62

Panzar, J. 7, 20, 26, 29
Pareto optimality 4
Paris
    Conference (1972) 151
    Convention (1883) 3
    Court of Appeals 163, 164
Pennell, Justice 82
Pennington case 57
Perfectly contestable markets (PCM) 20
Perry, M.K. 17
Petrochemical producers 221
Petrol supply case 121
Pharmacists, in France 168, 170–171
    Clin Midi 171, 172
    distribution of cosmetics by 177, 178
Philadelphia National Bank case 61, 62
Philip Morris case 234–235
Philips, L. 15
Pindyck, R.S. 30
Posner, R.A. 13, 51, 52, 56, 203
Poserian dichotomy 10
    LP approach 27
Postal franking machines case 124

Price freedom and competition, 1986 ordinance 156–162
Proctor and Gamble 121
Product-specific market power 23
Profit 34–35
    evaluation of performance 122
Public policy-makers, in France 146
Purchasing, exclusive 225
PVC and low-density polyethylene 221
Pynchon, T. 16

Quinn, J. 98

Raising rival's costs 56
Raleigh's refusal to supply 143, 144
Ravenscraft, D.J. 35
R & D
    investments 24, 112
    and new industries 216–219
Reagan Administration 8, 47
Realemon case 55
Recommended resale prices report 141
Reinganum, J. 11, 17, 23
Resale Prices Act (1964) 106, 136, 141
Resale Price Maintenance (RPM) 106, 142
    and discounts to retailers 141–142
'Restraints of trade' 250
Restrictive-distribution systems, and competition 174
Restrictive Trade Practices Act 106, 107, 108, 109, 136
    Commissioner (RTPC) 80
    Court 106
    Policy Review 114
Retailers, discounts to 141–142
Ricardian rents 12
Richardson, G.B. 113
Ries, W. 197
Robinson, J. 26
    Patman Act 195
Roosevelt, Theodore, Administration 45
Rosenbluth 89
Rothmans International 235
Rowntree Mackintosh, acquisition by Nestlé 185
Rule-making for competition policy 250–252
'Rule of reason' 251
Rules or discretionary authority 248–250

# INDEX

Salinger, M.A. 9
Saloner, (?) 19
Salop, S.C. 52, 56
Samuels, W. 1, 6
Sawyer, M. 127
Scandinavian Airlines 236
Scheffman, D.T. 37, 52, 56
Scherer, F. 16, 31, 35, 55, 57, 63
Schmalensee, R. 12, 13, 18, 21, 31, 33, 34, 36, 55
Schmidt, I. 197
Schmidtchen, D. 203
Schumpeter thesis 11, 24
Schwartz, M. 20
Schwinn v. Von's Grocery case 46, 47, 68, 69
Scott, J.T. 11, 24
Selective-distribution agreement 176, 225–227
Servan Schreiber, J.J. 147
Shapiro, C. 14, 34, 39
Sharp Electronics decision 69
Sharpe, T.A.E. 143
Shaw, R.W. 124
Shepherd, W.G. 17, 19, 20, 22, 31, 230
Sherman, R. 59
Sherman Antitrust Act (1890) 43, 44, 50, 53, 54, 84, 92, 190, 191, 192, 193, 194, 195, 250
Sherwin, R. 37
Simpson, P. 124
Singh, A. 131
Single price, assumption of 14–16
Slade, M. 2, 37, 213
SMEs (small and medium-sized enterprises) 211, 121–222
Smiley, R.H. 14
Smith, Adam 3
Smith, W.F. 64
Spence, A.M. 52
Spiller, P. 37
Spontex, acquisition of by 3M 185, 186, 188
Standard oil Company 53
State aids, and cooperative agreements 214–215
Static deadweight cost of monopoly 12–18
 assumption of a single price 14–16
 assumption of fixed quality 16
 marginal cost pricing 17

status of monopoly rents 17–18
Status of monopoly rents 17–18
Steel-makers, French, and Italian importers 184
Stigler, J. 20, 36
Stiglitz, J. 20
Stocking, G.W. 52
Stykolt, S. 75
Sugar
 Atlantic case 83, 84, 86
 BC case 91, 94, 95
Sullivan, L.A. 8
Supercentrales d'achat 156
Supply, refusal to 142–144
Sykes, A.O. 34, 35
Sylvania case 47, 49, 69
Synthetic resins 188

Takeover mechanism 131
Tampons profits case 124
Technical or economic progress, improving 241
Telser, L. 70
Thomas, L.G. 21, 22
Thorn Electrical Industries, and Locatel 186
Thread manufacturers 167
Tie-in sales 136–138
Tirole, J. 15, 16, 17, 19, 39
Treaty of Rome 77, 116, 206, 207, 248, 250
 Article 85 209–210, 211, 220, 223, 233, 237, 238, 243, 251
 Article 86 228, 231, 232, 233, 234, 237, 238, 243, 251
 Regulation 67/67 224
 Regulation 1984/83 225
Turner, D.F. 57, 60, 65

Undertakers' services 179
Unilever 121, 139, 213
United Biscuits case 129
United Brands case 229, 240
United Kingdom
 competition policy 104–143
 horizontal collusive behaviour 109–118
 legislation 109–112
 effectiveness of 112–115
 1988 Green paper 115–117
 mergers 126–136

policy 127–132
proposals for reform 133–135
monopoly and the abuse of market
    power 118–126
  legislation 118–119
  definition of monopoly 119–120
  Commission's assessment of
    monopolies 120–121
  problems of assessment 121–123
  remedies 123–124
  effectiveness of 124–126
vertical relationships 136–144
  ties 136–138
  exclusive dealing 139
  franchising 140–141
  RPM and discounts to
    retailers 141–142
  refusal to supply 142–144
United States 247
  Department of Justice Merger
    Guidelines 29, 37, 46, 64, 70,
    197
  Federal Trade commission 248, 249
  policy
    antitrust 43–72
    competition
      comparison with West
        Germany 189–205
      essential rules for 250
    towards mergers 5

Vanneen, W.L. 101
Varian, H.R. 15
Vertical arrangements, between
    firms 223–228
  discounts 227–228
  exclusive distribution 223–224
  exclusive purchasing 225
  selective distribution and
    franchising 225–227
Vertical contractual relationships 67–71,
    136–145
  ties 136–138
  exclusive dealing 139
  franchising 140–141

resale price maintenance and discounts
    to retailers 141–142
  refusal to supply 142–144
Vertical restraints 100–102
Vertical restrictive agreements 173–177
Vickers, J. 10, 135, 230–236
von Hayek, F. 203

Wall, D. 37
Water distribution systems 179
Waverman, L. 73–103, 246–252
Webber, F. 77, 84
Weiss, L.W. 59
Wetston, H. 100
West German
  Federal Cartel Office 117, 191–192,
    194–195, 196
  competition policy, comparison with
    United States 189–205
    hitorical experiences 189–190
    basic philosophies 190–191
  Supreme Court 193, 202
  Monopolies Commission 199, 200
  Federal Court of Appeal 199
  critics of policy
    Association of Manufacturers
      (Bundesverband der Deutschen
      industrie) 202
Whinston, M.D. 11
White, L.J. 44
  salt case 124
Wiley, J.S. 51
Williamson, O.E. 56, 57, 59, 60, 99, 211
Willig, R.D. 7, 20, 25, 26, 29, 31, 33,
    34, 35, 36, 62
Winter, R.A. 100
Worsinger, L.P. 68
Württembergische metallwarenfabrik
    (WMF) and Rheinmetall 202

Xidex Corporation case 59

Yamey, B.S. 142
Yarn Spinners case 112
Yarrow, G. 230

# FUNDAMENTALS OF PURE AND APPLIED ECONOMICS

## SECTIONS AND EDITORS

**BALANCE OF PAYMENTS AND INTERNATIONAL FINANCE**
W. Branson, Princeton University

**DISTRIBUTION**
A. Atkinson, London School of Economics

**ECONOMIC DEVELOPMENT STUDIES**
S. Chakravarty, Delhi School of Economics

**ECONOMIC HISTORY**
P. David, Stanford University, and M. Lévy-Leboyer, Université Paris X

**ECONOMIC SYSTEMS**
J.M. Montias, Yale University

**ECONOMICS OF HEALTH, EDUCATION, POVERTY AND CRIME**
V. Fuchs, Stanford University

**ECONOMICS OF THE HOUSEHOLD AND INDIVIDUAL BEHAVIOR**
J. Muellbauer, University of Oxford

**ECONOMICS OF TECHNOLOGICAL CHANGE**
F. M. Scherer, Harvard University

**EVOLUTION OF ECONOMIC STRUCTURES, LONG-TERM MODELS, PLANNING POLICY, INTERNATIONAL ECONOMIC STRUCTURES**
W. Michalski, O.E.C.D., Paris

**EXPERIMENTAL ECONOMICS**
C. Plott, California Institute of Technology

**GOVERNMENT OWNERSHIP AND REGULATION OF ECONOMIC ACTIVITY**
E. Bailey, Carnegie-Mellon University, USA

**INTERNATIONAL ECONOMIC ISSUES**
B. Balassa, The World Bank

**INTERNATIONAL TRADE**
M. Kemp, University of New South Wales

**LABOR AND ECONOMICS**
F. Welch, University of California, Los Angeles, and J. Smith, The Rand Corporation

**MACROECONOMIC THEORY**
J. Grandmont, CEPREMAP, Paris

**MARXIAN ECONOMICS**
J. Roemer, University of California, Davis
**NATURAL RESOURCES AND ENVIRONMENTAL ECONOMICS**
C. Henry, Ecole Polytechnique, Paris
**ORGANIZATION THEORY AND ALLOCATION PROCESSES**
A. Postlewaite, University of Pennsylvania
**POLITICAL SCIENCE AND ECONOMICS**
J. Ferejohn, Stanford University
**PROGRAMMING METHODS IN ECONOMICS**
M. Balinski, Ecole Polytechnique, Paris
**PUBLIC EXPENDITURES**
P. Dasgupta, University of Cambridge
**REGIONAL AND URBAN ECONOMICS**
R. Arnott, Queen's University, Canada
**SOCIAL CHOICE THEORY**
A. Sen, Harvard University
**TAXES**
R. Guesnerie, Ecole des Hautes Etudes en Sciences Sociales, Paris
**THEORY OF THE FIRM AND INDUSTRIAL ORGANIZATION**
A. Jacquemin, Université Catholique de Louvain

# FUNDAMENTALS OF PURE AND APPLIED ECONOMICS
## PUBLISHED TITLES

Volume 1 (International Trade Section)
GAME THEORY IN INTERNATIONAL ECONOMICS
by John McMillan

Volume 2 (Marxian Economics Section)
MONEY, ACCUMULATION AND CRISIS
BY Duncan K. Foley

Volume 3 (Theory of the Firm and Industrial Organization Section)
DYNAMIC MODELS OF OLIGOPOLY
by Drew Fudenberg and Jean Tirole

Volume 4 (Marxian Economics Section)
VALUE, EXPLOITATION AND CLASS
by John E. Roemer

Volume 5 (Regional and Urban Economics Section)
LOCATION THEORY
by Jean Jaskold Gabszewicz and Jacques-François Thisse, Masahisa Fujita, and Urs Schweizer

Volume 6 (Political Science and Economics Section)
MODELS OF IMPERFECT INFORMATION IN POLITICS
by Randall L. Calvert

Volume 7 (Marxian Economics Section)
CAPITALIST IMPERIALISM, CRISIS AND THE STATE
by John Willoughby

Volume 8 (Marxian Economics Section)
MARXISM AND "REALLY EXISTING SOCIALISM"
by Alec Nove

Volume 9 (Economic Systems Section)
THE NONPROFIT ENTERPRISE IN MARKET ECONOMIES
by Estelle James and Susan Rose-Ackerman

Volume 10 (Regional and Urban Economics Section)
URBAN PUBLIC FINANCE
by David E. Wildasin

Volume 11 (Regional and Urban Economics Section)
URBAN DYNAMICS AND URBAN EXTERNALITIES
by Takahiro Miyao and Yoshitsugu Kanemoto

Volume 12 (Marxian Economics Section)
DEVELOPMENT AND MODES OF PRODUCTION IN MARXIAN ECONOMICS: A CRITICAL EVALUATION
by Alan Richards

Volume 13 (Economics of Technological Change Section)
TECHNOLOGICAL CHANGE AND PRODUCTIVITY GROWTH
by Albert N. Link

Volume 14 (Economic Systems Section)
ECONOMICS OF COOPERATION AND THE LABOR-MANAGED ECONOMY
by John P. Bonin and Louis Putterman

Volume 15 (International Trade Section)
UNCERTAINTY AND THE THEORY OF INTERNATIONAL TRADE
by Earl L. Grinols

Volume 16 (Theory of the Firm and Industrial Organization Section)
THE CORPORATION: GROWTH, DIVERSIFICATION AND MERGERS
by Dennis C. Mueller

Volume 17 (Economics of Technological Change Section)
MARKET STRUCTURE AND TECHNOLOGICAL CHANGE
by William L. Baldwin and John T. Scott

Volume 18 (Social Choice Theory Section)
INTERPROFILE CONDITIONS AND IMPOSSIBILITY
by Peter C. Fishburn

Volume 19 (Macroeconomic Theory Section)
WAGE AND EMPLOYMENT PATTERNS IN LABOR CONTRACTS:
MICROFOUNDATIONS AND MACROECONOMIC IMPLICATIONS
by Russell W. Cooper

Volume 20 (Government Ownership and Regulation of Economic Activity Section)
DESIGNING REGULATORY POLICY WITH LIMITED INFORMATION
by David Besanko and David E. M. Sappington

Volume 21 (Economics of Technological Change Section)
THE ROLE OF DEMAND AND SUPPLY IN THE GENERATION AND DIFFUSION OF TECHNICAL CHANGE
by Colin G. Thirtle and Vernon W. Ruttan

Volume 22 (Regional and Urban Economics Section)
SYSTEMS OF CITIES AND FACILITY LOCATION
by Pierre Hansen, Martine Labbé, Dominique Peeters and Jacques-François Thisse, and J. Vernon Henderson

Volume 23 (International Trade Section)
DISEQUILIBRIUM TRADE THEORIES
by Motoshige Itoh and Takashi Negishi

Volume 24 (Balance of Payments and International Finance Section)
THE EMPIRICAL EVIDENCE ON THE EFFICIENCY OF FORWARD AND FUTURES FOREIGN EXCHANGE MARKETS
by Robert J. Hodrick

**Volume 25** (Economic Systems Section)
THE COMPARATIVE ECONOMICS OF RESEARCH DEVELOPMENT AND INNOVATION IN EAST AND WEST: A SURVEY
by Philip Hanson and Keith Pavitt

**Volume 26** (Regional and Urban Economics Section)
MODELING IN URBAN AND REGIONAL ECONOMICS
By Alex Anas

**Volume 27** (Economic Systems Section)
FOREIGN TRADE IN THE CENTRALLY PLANNED ECONOMY
by Thomas A. Wolf

**Volume 28** (Theory of the Firm and Industrial Organization Section)
MARKET STRUCTURE AND PERFORMANCE – THE EMPIRICAL RESEARCH
by John S. Cubbin

**Volume 29** (Economic Development Studies Section)
STABILIZATION AND GROWTH IN DEVELOPING COUNTRIES: A STRUCTURALIST APPROACH
by Lance Taylor

**Volume 30** (Economics of Technological Change Section)
THE ECONOMICS OF THE PATENT SYSTEM
by Erich Kaufer

**Volume 31** (Regional and Urban Economics Section)
THE ECONOMICS OF HOUSING MARKETS
by Richard F. Muth and Allen C. Goodman

**Volume 32** (International Trade Section)
THE POLITICAL ECONOMY OF PROTECTION
by Arye L. Hillman

**Volume 33** (Natural Resources and Environmental Economics Section)
NON-RENEWABLE RESOURCES EXTRACTION PROGRAMS AND MARKETS
by John M. Hartwick

**Volume 34** (Government Ownership and Regulation of Economic Activity Section)
A PRIMER ON ENVIRONMENTAL POLICY DESIGN
by Robert W. Hahn

**Volume 35** (Economics of Technological Change Section)
TWENTY-FIVE CENTURIES OF TECHNOLOGICAL CHANGE
by Joel Mokyr

**Volume 36** (Government Ownership and Regulation of Economic Activity Section)
PUBLIC ENTERPRISE IN MONOPOLISTIC AND OLIGOPOLISTIC INDUSTRIES
by Ingo Vogelsang

**Volume 37** (Economic Development Studies Section)
**AGRARIAN STRUCTURE AND ECONOMIC UNDERDEVELOPMENT**
by Kaushik Basu

**Volume 38** (Macroeconomic Theory Section)
**MACROECONOMIC POLICY, CREDIBILITY AND POLITICS**
by Torsten Persson and Guido Tabellini

**Volume 39** (Economic History Section)
**TOPOLOGY OF INDUSTRIALIZATION PROCESSES IN THE NINETEENTH CENTURY**
by Sidney Pollard

**Volume 40** (Marxian Economics Section)
**THE STATE AND THE ECONOMY UNDER CAPITALISM**
by Adam Przeworski

**Volume 41** (Theory of the Firm and Industrial Organization Section)
**BARRIERS TO ENTRY AND STRATEGIC COMPETITION**
by Paul Geroski, Richard J. Gilbert, and Alexis Jacquemin

**Volume 42** (Macroeconomic Theory Section)
**REDUCED FORMS OF RATIONAL EXPECTATIONS MODELS**
by Laurence Broze, Christian Gouriéroux, and Ariane Szafarz

**Volume 43** (Theory of the Firm and Industrial Organization Section)
**COMPETITION POLICY IN EUROPE AND NORTH AMERICA: ECONOMIC ISSUES AND INSTITUTIONS**
by W. S. Comanor, K. George, A. Jacquemin, F. Jenny, E. Kantzenbach, J. A. Ordover, and L. Waverman

Further titles in preparation
ISSN: 0191-1708